Development in Practice
Series Editor: Brian Pratt

Each title in the *Development in Practice Books* series offers a focused overview of practice-relevant analysis, experience, and research on key topics in development.

Participatory Research and Gender Analysis
New Approaches
Edited by Nina Lilja, John Dixon, and Deborah Eade

Achieving Education for All through public–private partnerships?
Non-state provision of education in developing countries
Edited by Pauline Rose

Global Food Price Shocks and Poor People
Themes and Case Studies
Edited by Marc J. Cohen and Melinda Smale

Child Protection in Development
Edited by Michael Bourdillon and William Myers

Child Protection in Development

Edited by
Michael Bourdillon and William Myers

Routledge
Taylor & Francis Group
LONDON AND NEW YORK

Citation Information

The chapters in this book were originally published in *Development in Practice*, volume 22, issue 4 (June 2012). When citing this material, please use the original page numbering for each article, as follows:

Preface
Jo Boyden pp. 436

Chapter 1
Introduction: development, children, and protection
William Myers and Michael Bourdillon
Development in Practice, volume 22, issue 4 (June 2012) pp. 437-447

Chapter 2
Beyond war: 'suffering' among displaced Congolese children in Dar es Salaam
Gillian Mann
Development in Practice, volume 22, issue 4 (June 2012) pp. 448-459

Chapter 3
Protecting children from trafficking in Benin: in need of politics and participation
Neil Howard
Development in Practice, volume 22, issue 4 (June 2012) pp. 460-472

Chapter 4
The spatialisation of child protection: notes from the occupied Palestinian territory
Jason Hart
Development in Practice, volume 22, issue 4 (June 2012) pp. 473-485

Chapter 5
Following the law, but losing the spirit of child protection in Kenya
Elizabeth Cooper
Development in Practice, volume 22, issue 4 (June 2012) pp. 486-497

Chapter 6
Children's migration for work in Bangladesh: the policy implications of intra-household relations
Karin Heissler
Development in Practice, volume 22, issue 4 (June 2012) pp. 498-509

Chapter 7
Child protection and harmful traditional practices: female early marriage and genital modification in Ethiopia
Jo Boyden, Alula Pankhurst and Yisak Tafere
Development in Practice, volume 22, issue 4 (June 2012) pp. 510-522

Chapter 8
Global priorities against local context: protecting Bhutanese refugee children in Nepal
Rosalind Evans and Rachel Mayer
Development in Practice, volume 22, issue 4 (June 2012) pp. 523-535

Chapter 9
Rethinking orphanhood and vulnerability in Ethiopia
Gina Crivello and Nardos Chuta
Development in Practice, volume 22, issue 4 (June 2012) pp. 536-548

Chapter 10
Children's responses to risk in agricultural work in Andhra Pradesh, India
Virginia Morrow and Uma Vennam
Development in Practice, volume 22, issue 4 (June 2012) pp. 549-561

Chapter 11
'Risky lives': risk and protection for children growing-up in poverty
Kirrily Pells
Development in Practice, volume 22, issue 4 (June 2012) pp. 562-573

Chapter 12
Action research exploring information communication technologies (ICT) and child protection in Thailand
Philip H. Cook, Cheryl Heykoop, Athapol Anuntavoraskul, and Jutarat Vibulphol
Development in Practice, volume 22, issue 4 (June 2012) pp. 574-587

Chapter 13
Child protection: a role for conditional cash transfer programmes?
Natalia Streuli
Development in Practice, volume 22, issue 4 (June 2012) pp. 588-599

Chapter 14
Listening to Iraqi refugee children in Jordan, but then what? Exploring the impact of participatory research with children
Martha Nelems and Vanessa Currie
Development in Practice, volume 22, issue 4 (June 2012) pp. 600-612

Chapter 15
Concluding reflections: how might we really protect children?
William Myers and Michael Bourdillon
Development in Practice, volume 22, issue 4 (June 2012) pp. 613-620

Previously published *Development in Practice* Readers
Series Editor: Deborah Eade

Rawat Publishers and Practical Action Publishing, New Delhi and Rugby

Deconstructing Development Discourse: Buzzwords and Fuzzwords (2010)
Edited by Andrea Cornwall and Deborah Eade

Kumarian Press, Bloomfield, CT

Development and Humanitarianism: Practical Issues (2007)
Edited by Deborah Eade and Tony Vaux

Development and the Private Sector: Consuming Interests (2006)
Edited by Deborah Eade and John Sayer

Development NGOs and Labor Unions: Terms of Engagement (2005)
Edited by Deborah Eade and Alan Leather

Oxfam, Oxford

Development, Women, and War: Feminist Perspectives (2004)
Edited by Haleh Afshar and Deborah Eade

Development Methods and Approaches: Critical Reflections (2003)
Edited by Deborah Eade

Development and the Learning Organisation (2003) (in association with the Institute of Development Studies and Oxfam America)
Edited by Laura Roper, Jethro Pettit, and Deborah Eade

Development and Culture (2002) (in association with World Faiths Dialogue on Development)
Edited by Deborah Eade

Development and Cities (2002) (in association with UNRISD)
Edited by David Westendorff and Deborah Eade

Development and Agroforestry: Scaling Up the Impact of Research (2002) (in association with the World Forestry Centre, CGIAR)
Edited by Steven Franzel, Peter Cooper, Glenn L. Denning, and Deborah Eade

Development and Advocacy (2002)
Edited by Deborah Eade

Debating Development: NGOs in the Future (2001) (in association with Oxfam International)*
Edited by Deborah Eade and Ernst Ligteringen

Development, NGOs, and Civil Society (2000)*
Edited by Deborah Eade

Development and Management: Experiences of Value-Based Conflict (2000) (in association with The Open University)
Edited by Deborah Eade, Tom Hewitt, and Hazel Johnson

Development and Social Action (1999)
Edited by Deborah Eade

Development with Women (1999)
Edited by Deborah Eade

Development and Rights (1998)*
Edited by Deborah Eade

Development for Health (1997)
Edited by Deborah Eade

Development and Patronage (1997)*
Edited by Deborah Eade

Development and Social Diversity (1996) *
Edited by Deborah Eade

Development in States of War (1996)*
Edited by Deborah Eade

* Also published in Spanish translation by Intermón, Barcelona.

Notes on Contributors

Jo Boyden has been Director of Young Lives since 2005. Her research has mainly focused on child labour, children and political violence, and childhood poverty – particularly in bringing together academics, practitioners and policymakers to develop effective models and methods for supporting children, their families and their communities in situations of adversity.

Michael F.C. Bourdillon is Professor Emeritus in the Department of Sociology, University of Zimbabwe.

Nardos Chuta is an Ethiopian qualitative researcher with Young Lives, based in Addis Ababa. Her research has focused on children's work and migration and the conceptualisation of childhoods in Ethiopia.

Philip H. Cook is Executive Director of the International Institute for Child Rights and Development (IICRD), Centre for Global Studies, University of Victoria, Canada

Elizabeth Cooper is a postdoctoral research fellow at the School for International Studies at Simon Fraser University in Vancouver, Canada. She has a doctorate in social anthropology from the University of Oxford and has conducted research concerning poverty, orphanhood, land rights, forced migration, and child protection in different parts of sub-Saharan Africa.

Gina Crivello is an anthropologist and Research Officer with Young Lives, based in Oxford. Since 2009, she has co-managed a study on the role of research in improving policy and practice in the area of child protection in Ethiopia and India.

Vanessa Currie is a consultant in children's well-being, protection and participation, using holistic approaches to reimagine community development. Her work focuses on engaging young people, their families and communities in participatory action research, and monitoring and evaluation processes that draw community wisdom forward in partnership with governments, NGOS and civil society. Working in the global south as well as with Indigenous communities in Canada, Vanessa leads the development of innovative culturally appropriate and child and family centred systems. Community-based child protection mechanisms that reinforce communities' capacity as agents to support young people are an area of special interest. Vanessa serves as an Associate with the International Institute for Child Rights and Development at the Centre for Global Studies at the University of Victoria and as an Associate with Columbia University's Columbia Group for Children in Adversity.

Rosalind Evans studied at the University of Oxford from 2005-09, achieving an MSc in Forced Migration and a doctorate in international development studies. Her research focused on children's participation programmes in refugee camps in Nepal. She currently works as a practitioner with young refugees in the UK and overseas.

Jason Hart is a Lecturer in the Department of Social and Policy Studies, University of Bath, and a Research Associate at the Refugee Studies Centre, University of Oxford. Much of Jason's work has explored the experience of young people on the margins of society and the global economy: relating these to the values, assumptions and approaches of agencies seeking to provide support. Children living in situations of political violence and forced displacement have been a particular focus of much of his research and writing. His principal area of interest is the Middle East, particularly Israel and occupied Palestinian territories, and Jordan.

Karin Heissler is a child protection specialist with UNICEF. She has a doctorate in Development Studies from the University of Oxford. Her research interests include intra-household dynamics around choice, children and work, and migration.

Cheryl Heykoop is a Project Officer at the International Institute for Child Rights and Development (IICRD), Centre for Global Studies, University of Victoria, Canada

Neil Howard is a research student in the Oxford Department of International Development at Oxford University. His research interests include international aid policy and praxis; the politics of (forced) migration; human trafficking and anti-trafficking policy; child labour and migration; childhood and family studies and the interaction between the different levels of the policy system.

Gillian Mann, PhD, is a freelance child protection consultant with a background in education and anthropology. In addition to her academic work, she has conducted research, evaluation, and training for a broad range of agencies on issues related to children's experiences of forced migration, displacement, family separation, HIV/AIDS, and poverty.

Rachel Anne Kiernan Mayer holds an MSc in Refugee and Forced Migration Studies from the University of Oxford and a first-class degree in International Relations and Anthropology from the University of Sussex. Her research focuses on young refugees in the context of refugee resettlement. She works with unaccompanied minors in Cairo.

William E. Myers is an educator who has worked with UNICEF and ILO, and is an associate in the Department of Human and Community Development, University of California, Davis, USA.

Virginia Morrow is Senior Research Officer at Young Lives UK. In addition to writing about children's involvement in agricultural work in Andhra Pradesh (with Uma Vennam), based on qualitative data, she also co-manages a sub-study funded by the Oak Foundation, exploring the role of research in policy and practice in relation to child labour in India. The working paper she wrote on research ethics for Young Lives has been widely cited, and a summary has been included as a case study in the ESRC's revised Framework for Research Ethics (2010).

Martha Nelems is a long-term advocate of children's rights, and an experienced negotiator, policy analyst, researcher, project manager, evaluator, and facilitator. She has worked as senior policy analyst in children's rights at the Canadian International Development Agency, and as a consultant with UNICEF, the African Development Bank, international and national non-governmental organisations and various community-based groups. She is especially interested in helping organisations translate their child protection policies into action, and in promoting children's participation within efforts to strengthen national and community based child protection systems. She is an associate of the Columbia Group for Children in Adversity at Columbia University and of the International Institute for Child Rights and Development at the University of Victoria. She currently lives in Tunis, Tunisia.

Alula Pankhurst is a social anthropologist and joined Young Lives as Ethiopia Country Director in November 2009. His role involves representing Young Lives and coordinating the work in Ethiopia. His research interests include community change; household types and cycles; well-being; and poverty dynamics.

Kirrily Pells completed a PhD and postdoctoral fellowship on children and young people in post-conflict situations with case studies of Rwanda and Bosnia. Kirrily has worked as a consultant for international organisations in the child protection field. She is currently a policy officer at Young Lives UK.

Natalia Streuli received her PhD in Childhood Studies from the Institute of Education, University of London, and was the Lead Researcher for the qualitative component of the Young Lives project in Peru.

Yisak Tafere is Lead Qualitative Researcher for Young Lives Ethiopia. He has an MA in Social Anthropology and BA in Philosophy from Addis Ababa University. He has led four rounds of qualitative research fieldwork in Ethiopia, in collaboration with colleagues in Oxford and Ethiopia. He has also presented papers at conferences, and authored working papers, book chapters, and journal articles related to Young Lives. His research interests include cultural constructions of childhood, well-being and transitions; risk, resilience, aspirations, and achievements of children and young people; social protection; and children's work and educational trajectories.

Uma Vennam is Lead Qualitative Researcher at Young Lives India. Uma has an MA in Social Work, specialising in Urban and Rural Community Development, from Tata Institute of Social Sciences Bombay and holds a PhD from the University of East Anglia, UK. She is currently a Professor of Social Work at SPMVV. She has been involved in various projects dealing with poverty alleviation, rural livelihoods, SLMF, poverty and HIV/AIDS, trafficking of women and children, child labour and AIDS Prevention Education Programme (APEP) for school children, working with international agencies including the World Bank, DFID, and UNICEF.

Dedication

This special issue of *Development in Practice* is dedicated to Natalia Streuli Wilson (1976–2012), who sadly passed away prior to its publication following a brave struggle against cancer. Natalia conducted her doctoral fieldwork in a Young Lives site in Peru and worked with Young Lives from 2008 until her death; initially in Oxford, and more recently in Lima, she had taken on the role of a lead qualitative researcher for Peru. Natalia was a highly valued and committed member of the team whose contribution will have lasting effect on our research and policy engagement. A cherished friend, she is sorely missed by everyone.

Preface

Jo Boyden

This special issue on child protection reflects a growing awareness among many researchers and practitioners of the sheer numbers of boys and girls in developing countries who continue to be exposed to poverty and other related adversities, despite recent concerted efforts to improve children's survival and well-being. There is evidence that too often our efforts to protect children are not as effective as we hope and, worse, there have been nagging suspicions that they are sometimes even misguided and counterproductive. It has gradually become clear that the challenge to understand why and how children become vulnerable flows into the challenge of understanding why and how our efforts to protect them succeed or fail. It is that question that this journal issue addresses.

The Oak Foundation was concerned to better understand some of the numerous risks confronted by children in developing countries, taking into account as far as possible boys' and girls' own perspectives, and to propose ways forward for policy: to this end, the Foundation was keen to initiate an exchange between researchers and practitioners by funding a workshop hosted by the Young Lives study in Oxford in May 2011. Independent scholars and Young Lives staff were brought together to share findings from their research with children in diverse contexts and consider the implications for policy. The contributors are all experienced researchers and practitioners who have worked for many years with children in developing countries. It soon became apparent in the workshop that participants shared a critical perspective on current policies and a consensus was reached on the need for new approaches. The articles presented here reflect this critical perspective and offer suggestions for reform based on empirical findings around a range of child protection concerns, including children's work, independent migration, family separation, early marriage, and military occupation. The introduction and concluding reflections, co-authored by Michael Bourdillon and William Myers, who also facilitated the workshop, draw out the key messages. Together, these contributions provide an important body of knowledge which is applicable to humanitarian and development policy and practice.

Young Lives' involvement in this work has been motivated by evidence emerging from the study that poverty is a key indicator for multiple risks in children. Building on a multidisciplinary perspective and using a mix of survey and qualitative data, the study is tracking 12,000 children, their households, and communities over 15 years, in Ethiopia, India (Andhra Pradesh), Peru, and Vietnam. The boys and girls in the sample are in two age groups, one born around 2000–2001 and the second, 1994–1995. They were drawn randomly from approximately 20 poor rural and urban sites in each country that were selected to reflect country diversity in terms of livelihoods, ethnicity, religion, and other social determinants.

For more information about Young Lives, visit www.younglives.org.uk. Longer versions of several of the articles in this special issue can be found on the Young Lives website (searchable at www.younglives.org.uk/our-publications).

Introduction: development, children, and protection

William Myers and Michael Bourdillon

This paper introduces the special issue. It first places the protection of children in the context of development studies. It goes on to outline current international trends in the protection of children and raises questions about them. Finally, it introduces the papers in the issue and highlights how they speak to the questions raised.

Introduction : développement, enfants et protection
Cet article constitue l'introduction du numéro spécial. Il situe d'abord la protection de l'enfance dans le contexte des études du développement. Il présente ensuite les tendances internationales actuelles dans la protection de l'enfance et soulève des questions à leur sujet. Il présente enfin les articles figurant dans le numéro et met en relief la manière dont ils traitent des questions soulevées.

Introdução: desenvolvimento, crianças e proteção
Este artigo introduz a edição especial. Primeiramente, ele insere a proteção das crianças no contexto dos estudos de desenvolvimento. O artigo, então, apresenta as tendências internacionais de hoje relativas à proteção das crianças e levanta questões sobre elas. Por fim, os artigos da edição são apresentados e destaca-se como eles abordam as questões levantadas.

Introducción: desarrollo, infancia y protección
El ensayo que introduce este número especial comienza situando la protección de la niñez en el contexto de los estudios de desarrollo. Continúa perfilando las tendencias actuales de la protección de la niñez en el ámbito internacional y plantea dudas sobre las mismas. El ensayo concluye con una semblanza de los artículos que componen este número y se centra en cómo éstos responden a las dudas planteadas.

Why should a journal on development studies support a special issue on protecting children? First, development requires attending to children, who in developing countries may comprise half the population or more. Their presence and status cannot be ignored, they have needs specific to their young age, and major national policies and resources are devoted to their needs. Second, they will control the future, perhaps even sooner than many expect (as recent events in the Middle East and elsewhere might suggest), and long-term development policy must address their protection and preparation to assume responsibility for society not many years from now. Third, as human development assumes increasing prominence in development thinking, how to protect and rear children becomes a central concern that strongly influences the parameters of possibility. Moreover, the protection of children (the term 'child protection' has problematic connotations that we shall discuss shortly) cannot be understood without considering how children are affected by development policy more broadly. The directions in which both world events and development concerns are moving suggest that international ideas, policies, and actions for the protection of children need to be seriously reconsidered. This journal issue is intended to provoke and inform that rethinking.

The first four papers in this collection provide cases in which children have been seriously failed by current child protection policy and practice. In this introduction, we look at the background and context of current child protection to help understand why and how such failure becomes possible, and open up questions that the papers in this collection will help to answer. In the final paper in this issue, we reflect on how these questions have in fact been answered.

Protecting children in the context of development

Promotion of the well-being and protection of infants and young children is one of the greatest successes of international development. It has occurred through both general economic and social development assistance, and special policies and programmes targeted specifically to children. It has long been understood that the greatest threats to children in poor countries come from generalised conditions of poverty and violence that affect society as a whole, and that their well-being ultimately depends on economic and social development reducing these threats to everyone. Raising income levels, provision of public health and sanitation infrastructure, and the establishment of effective order and security from armed conflict and crime are examples of general development progress that have been massively beneficial for children. For the world as a whole, and despite many areas of challenge yet to be met, children are today far less likely to succumb to conditions of poverty or violence than was the case only decades ago. In the words of a prominent social scientist surveying this progress, especially in the reduction of violence, '*It is premature to say that the kids are all right, but they are certainly better off than they used to be*' (Pinker 2011: 442).

Development assistance has also supported interventions targeted at children specifically. The first priority has been protection against early death, and a package of interacting actions disseminated worldwide has helped cut infant and child mortality to less than half what it was only a few decades ago. Equally impressive reductions in maternal mortality have ensured that more children will grow up secure in the loving care of their mothers. Improved early childhood nutrition has both reduced exposure of children to debilitating illness and ensured children's proper physical and mental development. That achievement is primarily the result of expanding and upgrading the delivery of key public sector services, ranging from the organisation of well-baby clinics to early childhood education and massive vaccination campaigns. While most of the effort and resources come from national governments, bilateral and multilateral assistance has been essential to reaching these goals of infant and early childhood protection in the poorer countries. Even strong critics of international assistance in general tend to agree that these programmes protecting the survival and development of infants and young children are highly successful where they are fully implemented.

Two fundamental characteristics of programmes protecting children's life and development have underpinned their success. First, they have been rigorously empirical, based on quality research for understanding the problems to be addressed and the development and testing of solutions. Studies start with field situation analyses to ensure that threats to children are properly perceived and understood in context, continue through systematic monitoring of children, and follow through to the evaluation of the effectiveness of interventions as determined by their measured impact on the children involved. Using empirical evidence, policies and activities have been held accountable for their results. Much attention has been devoted to understanding the importance of social and economic context for children at risk, and indeed the Young Lives project responsible for this journal issue, and much of the research reported in it, represents precisely such an investment by the UK Department For International Development (DFID), the Netherlands Ministry of Foreign Affairs, the Bernard van Leer Foundation, and the Oak Foundation.[1]

Second, successful programmes to protect young children's lives and development have established delivery structures that engage, promote, and support families and communities to protect their own children. Even technologies introduced from the outside, such as rehydration salts and vaccination, have been delivered through cooperation with community structure and participation. The combination of field research and community engagement ensures that even technology based in universal physical and mental characteristics is contextualised as necessary to successfully reach most children. Solid grounding in empirical facts and community engagement are in these activities recognised pillars of good practice for protecting children, just as they are for meeting other development objectives.

However, this rigorously empirical and community-engaged approach to protecting children loses its dominance after early childhood. International policies and programmes for the protection of older children seem to have diluted, or even replaced, an empirical approach with a normative paradigm basing protective policies and programmes primarily on international law and political agreement. Whereas 'science' could be considered the reigning value of international protection policies for young children, for older children it is 'law'. It is highly questionable whether this shift has been good for children.

Comparing protective approaches for younger and older children raises a crucial question for development policy and practice: why are the rigorously empirical and community-responsive values that drive successful international action to protect the lives and development of young children not continued and extended in policies and programmes to protect older children? To put this question another way, what justifies the change away from a successful empirical and community-oriented approach for young children to one that is more political and less factually based – and therefore less accountable – in the protection of older children? Another question logically follows: is it good development practice to replace an approach empirically accountable for its results in children's lives with one that is not?

These questions are at this moment especially important, for the policy field commonly known as 'child protection' shows signs of moving toward a legalistic framework, prioritising adherence to universalised legal instruments and away from interventions based on empirical evidence and community responsiveness. The worry is that imposing international norms decided on the basis of distant power and influence rather than on the circumstances of children's lives threatens to be a move away from accountability for results in children.

Child protection trends and issues

There are two current trends in the extension of child protection internationally, and they sometimes conflict. Both try to make children's rights and protection more effective. One approach is 'universalisation', and it draws primarily on international law, legal theory, and the politics surrounding these. Its advocates hold that children's rights and protection are best served by globalising and standardising legal norms, codified and observed more or less equally everywhere. It establishes government-led, bureaucratic structures to enforce laws and principles. The other approach is 'contextualisation'. Based largely on findings of empirical social science and experience in community development, it adapts the implementation of principles such as human rights to the specific needs and characteristics of particular places and cultures. It connects with culture and organises families, communities, and societies to generate from their own values and experience understanding and practices consistent with the rights and protection of children.

Universalisation

This is very much a top-down approach. Governmental and inter-governmental actions tend to favour this approach, which works at the political level. Sometimes elaborate processes involve consulting experts and authorities from all over the world to formulate laws and implement rules that can apply everywhere. The inter-NGO Child Protection Workgroup, catering primarily for children in emergency, follows this approach as does the Protection Project in Johns Hopkins University, which is drafting a 'Child Protection Model Law',[2] which claims to offer '*best legal practices*', the objective of which is to '*protect children's best interests from any negative effect*'. These attempts are based on practices in high-income countries that are assumed to be technical and moral standards for all to follow. They show no concern to place the protection of children in the contexts of the communities in which they live, and consequently pay no attention to the needs of children in specific contexts, nor to outcomes for children. Although they affirm children's right to a say in decisions that affect them, these decisions are clearly determined by adults from outside the communities of the children concerned. Proposed research does not include outcomes of policies for children.

An example of this approach would be enforcement of a general minimum age for employment, which has never been seriously assessed in terms of outcomes for children (Bourdillon *et al.* 2010). It derives from the historical development of the notion of child protection in relatively high-income societies in response to the abuse of children, often children of the poor. This was particularly evident in responses to the exploitation of children in mines and factories early in the industrial revolution, and resulted in top-down legislation on how children may be treated. There have been considerable achievements, with states throughout the world taking responsibility for dealing with the gross and mass abuse of children.

In the twentieth century, the development of social work services has helped to control the abuse of children in their own homes. For some social workers, the main focus for policy and reform is to improve methods to catch those unprotected children who slip through the net, and perhaps to avoid intrusions that are disruptive to children and their families (e.g., Ferguson 2011).

But developed in such an environment, child protection is based on the assumption that only a minority of children need protection, and this need is judged in terms of a model of childhood that is achievable by the majority. It involves paying attention to particular cases in which the treatment of children does not meet the norms that are assumed to be achievable for all. This gives rise to two dominant reactive strategies. First, when cases are found in which children do not receive treatment according to accepted norms, they are treated on a case-by-case basis, on the assumption that society can supply what the child's family cannot or fail to supply – the dominant social work model. Second, when a number of such cases arise, preventative regulations are devised and assumed to provide protection to children generally (leading to a universalising approach to protection).

A major question raised about this social work model is whether or not, when children are at risk from things that face whole communities or societies, such as some form of social exclusion or poverty, they can be protected without looking at the broader issues. When a child is malnourished because the whole community lacks food, or when a child lacks schooling because there are no effective schools available, is trying to provide food for the individual child and a place in a distant school likely to be effective? Might not other approaches be more productive for children? Within the field of social work, traditional focus on individual children has now been seriously questioned in the context of deprived communities (e.g., Lonne *et al.* 2009).

Contextualisation

Partly in response to these questions, a second current trend has developed for extending the protection of children internationally, moving away from universalisation and standardisation and toward 'contextualisation'. This approach recognises the need for international human rights as guiding principles, but it seeks ways to implement them that reflect the situations, cultures, and communities to which they are applied. It looks for local cultural and community resources for children's rights and protection that can be mobilised. It is an approach that builds toward protection and rights as much from the bottom up as from the top down. This is a strategy used mostly by NGOs and communities who work directly with communities and children. This is in some respects a newer perspective on how to build child protection, one based intellectually in the findings and methods of empirical social science and socially in community dynamics, including the participation of children.

Whereas the technical foundation underlying the 'universalisation' approach is legal theory, what drives 'contextualisation' is a passion for evidence, the facts of people's lives and what affects them. Those facts typically come from two sources. One is structured research, such as academic cultural studies to find out how people live, situational analyses to diagnose problems, and evaluations to determine the reach and outcomes of policies and programmes. The other source is first-hand testimony by children, families, and communities regarding how they understand issues, what they think needs to be done, and how they assess the effectiveness of existing policies and programmes in their own lives. In recent times, research practice into child rights and protection issues has fused these sources in such a way that, increasingly, good field research provides for both facts *about* children and testimony *from* them. While this approach facilitates the adaptation of general principles and laws to the reality of particular populations, it has been open to the charge from some human rights quarters that in so doing it risks undermining and weakening the principles. Specially adapted rights, by this token, can be inferior rights. Those defending contextualisation respond that the quality of law need not suffer from creative application, and that in any event rights observed because they fit a social context are in the end more effective for children than are more inflexible positions that go unobserved because they seem too alien to reality.

Many now believe that arguments between 'universalisers' and 'contextualisers', although real and substantive, are in danger of being overblown and should be reconciled. Both schools of thought, it is suggested, should seek to move beyond their exclusiveness to fuse their contributions in ways that protect children and their rights more effectively. Some of the articles to follow suggest the need for this broader view.

Problems of risk orientation

There also are worries that much child protection characteristically reacts to things that have gone wrong, focusing only on the negative and the removal of risk. Focusing on removal of risk can result in a multitude of restrictions that often hinder children from doing things they wish to do and which they perceive to be in their short- and long-term interests. It has been noted that even in high-income societies, in which risk of kidnapping or other violence toward children is extremely low, parental worry and overprotection can restrict children so severely that they have little opportunity to learn how to manage risk or to learn responsibility (Pinker 2011: 443–7).

Such considerations highlight the limitations of considering protection exclusively in terms of eliminating physical harm or risk, and raise the fundamental question: in terms of the lives of children, what is protection ultimately for? Should protection also consider how meaningful and hopeful are children's lives and how opportunities might be opened – and kept open – for them? Meaningful lives and opportunities involve relations with others: to what extent should protection attend to a child's relations with family and community? Related to this is whether ideals and standards for childhood developed in high-income countries necessarily benefit children growing up in low-income communities and in very different cultures.

A further question has been raised as to whether 'child protection' is a useful category for policy and programmes. Does it perhaps encourage single-faceted responses, taking children out of development and indeed out of the societies in which they live? Does it encourage a harmful dichotomy between children and the adult world into which they are growing up? Is the notion of child protection too reactive, responding to threats, and insufficiently proactive in establishing conditions in which children thrive?

Morals and ethics

Child protection interventions are often framed in public discussion as moral imperatives, often expressed in terms of rights. Given the justifiable emotional reaction to many injustices suffered by children, this is understandable and may be inevitable. But under what conditions, and to what extent, are such moral judgments justified by the facts? Do they help protect children? The papers to follow raise such questions. Are even conventional measures widely promoted internationally always ethical? Are they sufficiently accountable to those who are to be protected?

Two currently prominent types of moral claims are particularly suspect. First is the assumption that moral claims – or rights – are based on intuitive values. The question raised by this assumption is whether the insistence on such moral claims or results too often results in ethnocentric bullying of the weak by the strong, a well known but widely ignored effect. Values-based claims, especially when universalised in international normative policies, raise the question of just whose values prevail and whose lose. How are winners and losers determined, and by whom?

The second type of claims now being questioned is political, claims based on agreements such as national laws and international conventions. Legitimacy and consensual authority would appear to give these the moral force of a social contract. Nevertheless, despite assertions of society-wide reach, doubts have been raised not only about whose values and interests they actually serve, but also about whether other issues intrude on decisions ostensibly made about and for children.

The values-based and agreement-based approaches are often brought together in legislating laws, guiding principles, formal rights, and other norms that establish duties and responsibilities. International conventions, such as the United Nations Convention on the Rights of the Child, expect to be successful by making everybody follow universalised rules. The advantage of this norm-based approach is bureaucratic simplicity – one need only follow the rules, on the assumption that doing so will produce the expected good outcomes. This is the stuff of law. But does this encourage consideration of whether the rules actually work for children? Some worry that when 'morality' shirks responsibility for outcomes, it is likely to degenerate into cheap moralism that is unethical and may itself become a danger to children. Consequences matter; and they matter so much that one school of thought, found in welfare economics and elsewhere, has considered them to be all that really matters. A result has been something of a philosophical divide between rules-based and results-based perspectives, and the professions attached to each. The articles to follow explore relationships between rules and consequences in the protection of children and their well-being.

Working from a series of field case studies to illustrate key points, this journal issue offers a critical view of current child protection policies, such as they are, from middle childhood through adolescence, in terms of their impact on children. It moves on to three case studies that offer ways in which intervention might be better protective of the well-being of children, whilst the final chapter reflects on lessons to be taken from the studies.

How things fall apart

The first four case studies illustrate how things can and do go badly wrong for children even when there are agencies present to protect them. Sometimes agencies focus on survival and protection from specific harm without considering the social context of meaningful lives and the importance of opening up possibilities for the future; so they simply do not see what are the fundamental problems faced by children.

The first paper of this section raises the fundamental question of what child protection is about. **Gillian Mann** shows how refugee children see social exclusion and lack of hope as their greatest problems, greater in some ways than the physical danger they have escaped from. These are social problems that affect their parents as well as the children. The children perceive the support offered to formally recognised refugees, combined as it is with restrictions, as a threat rather than a help (a point which is elaborated in Hart's discussion of spacialised protection).

The second paper illustrates how the application of universal standards and ideals can restrict the efforts and chances of young people who are trying to improve their lives. International standards of child protection have become dominant among local policymakers and agencies, with little attempt to find out from the people and the children what lies in children's interests. **Neil Howard** shows how attempts by children and their families to improve the children's opportunities through migration and work are frustrated by attempts to protect children from exploitation and trafficking. In spite of good intentions, intervention in fact damages children's opportunities. This study is supported by a growing literature showing how implementation of international standards against child labour, often couched in terms of rights of children, frequently hurt the children they are intended to protect.

In the following article, **Jason Hart** shows how spatial thinking dominates much intervention, and is inappropriate for the protection of Palestinian children in the territories occupied by Israel. Attempts to provide safe places for children do not help when the situation in which children live is fundamentally unsafe. Ideas derived from other places and other contexts are poor guides to appropriate intervention. In this situation, attempts to help children without considering fundamental injustices to the communities in which they live can at best provide superficial support. In a fuller report on this study, Hart and Lo Forte (2010) show that the problem is exacerbated when interests of international politics override the children's interests.

The importance of accountability in terms of outcomes for children becomes even clearer in the next paper, which further illustrates how adherence to formal rules is insufficient to protect children. Although Kenya has been praised in some arenas for its attention to child protection, **Elizabeth Cooper** shows that attention to bureaucratic procedures and priorities fails certain children who come to authorities in need of special care. Children's perspectives are largely ignored as officials try to streamline procedures for clearing cases, with disastrous consequences for some of the children.

In all the cases in this section, interests of different groups affect interpretations of children's needs and priorities in attending to them. The next section illustrates how local context can affect interpretations of child protection.

The importance of context

It has been established that strategies for child development can only work if attention is paid to the particular context of children (Dodge 2011). Context is equally important in protecting children. The three essays in this section illustrate how the contextual details of children's lives impact on them, and therefore on the kinds of protection that might be helpful.

In the first, **Karin Heissler** takes up the theme introduced by Neil Howard, but in a very different setting: she studies adolescents migrating for work in Bangladesh. Paying attention to the views of children and their supporting adults, she points to the interdependence of adults and children within families: children contribute to their families and expect support in return. Poverty and shocks affect children differently, depending on their position within the family. Even when the work of children is undesirable according to both local and international standards, it can have protective functions in household strategies to deal with shocks and poverty. Families and local communities, within a culture of shared values, provide some protection for children in the absence of outside help. For interventions to be effective in protecting such vulnerable children, conformity to standardised norms is not adequate: they need also to pay attention to the situation of children's families, to children's positions within their families, and to the local values that support them.

Jo Boyden, **Alulah Pankhurst**, and **Yisak Tafere** consider practices that are widely held to be harmful to girls – and indeed are demonstrably physically harmful in many cases; these practices nevertheless persist in certain Ethiopian communities. Although such customs are widely interpreted as arising from and supporting patriarchal authority, parents in the communities point to the fact that schools can be dangerous for unmarried and adolescent girls, and that early marriage can protect girls from sexual violence. The girls themselves sometimes demand initiation rites that can be harmful in order to save themselves from shame and to improve their chances of marriage. Simply to consider these practices as inhibiting schooling or causing physical harm is to ignore their importance in the emerging social lives of the children. Banning these practices cannot be regarding as protection if thereby the children are dislocated from their families and communities. Genuine protection from harmful practices requires sensitive involvement of the communities.

Rosalind Evans and **Rachel Mayer** continue the theme of dominant ideologies being imposed on people, without listening to the children. Here we find a tension between the priorities and values of the bureaucratic United Nations High Commission for Refugees (UNHCR) and the cultural values of a refugee Bhutanese community. Both the UNHCR and the local community want to protect the children, but they go about this protection in different ways. The communities provide some practical support for the children, which is not always appreciated by UNHCR officials: what appears to one group as protective can appear as risk to the other. Evans and Mayer explain this tension largely in terms of obtuse assumptions about childhood and family in the Global North that bear little relation to the experiences of the children in the camps.

These three papers illustrate the importance of understanding what people are doing and why before intervention can be genuinely protective. They highlight the fact the children are growing up in specific local communities; their opportunities for growth depend on integration into these communities. The most likely way to protect children effectively is to find out what local communities are already doing for them and to enhance these efforts rather than to introduce totally different strategies that are likely to compete with them. In particular, intervention should understand what children are doing for themselves in order to enhance their present and future prospects.

The effects of poverty

Notions of childhood from the Global North are least likely to be helpful in enabling children to respond to poverty in either the immediate or the longer term. Strategies developed locally, and especially by the children themselves, are more likely to bear on the problems they face than are assumptions from a very different experience of childhood. The effects of poverty have already been seen in some of the earlier papers, in particular Howard on migrant children; the next three papers bring this to the fore.

Before we consider them, we need to consider what poverty is. While a variety of indices for poverty have been promulgated, all have limited functions with respect to policy (Dercon 2012). There is the further problem that inequality can be a more serious social problem than poverty measured in absolute terms: to be denied benefits that are taken for granted by peers can be traumatic, while people may sometimes find satisfactory lives in communities where all have few resources. This factor has been well exposed in research in high-income countries, where high inequality is damaging for rich and poor alike (Wilkinson and Picket 2009). Data are sparse for the different situation of low-income countries, where many people lack such fundamental necessities as food and shelter. Nevertheless, some case studies suggest that a parallel situation applies: Mann's study shows social exclusion to be as important as material poverty, and data arising out of Young Lives research offers some support (for example, Crivello *et al.* 2011). In the papers of this section, poverty is not defined according to some determined index, but refers to areas in which the participants, including children, see their economic status as problematic.

Gina Crivello and **Nardos Chuta** study data from the Young Lives research programme on the effects of poverty in Ethiopia. Bureaucratic agencies focus on orphans as the most vulnerable children in need of protection, partly as a means of attracting financial support. Young Lives data, however, show poverty and economic shocks to be more relevant to vulnerability than is orphanhood, a conclusion supported by children's perceptions (and reinforcing the findings reported by Evans and Mayer among Bhutanese refugees). The paper shows that risk is related to relative poverty rather than to categories of vulnerability such as being orphaned, and suggests that interventions focussing on children that have been, or are likely to be, orphaned is likely to miss areas where the children most need support, and miss many of the most vulnerable children.

In the following paper, **Virginia Morrow** and **Uma Vennam** again take up the theme of work, presenting children's views of their work in agriculture, which diverge from an orthodoxy focussing on the elimination of child labour. In spite of the unpleasantness and hazards that children describe, and in spite of seeing little future for themselves in agriculture, children provide a number of reasons why they work, and fears of what might happen to their lives if they did not work. For some, work is part of their own and their families' strategies to stave off poverty. The children develop their own protective responses against some hazards in the workplace, and do not see a prohibition on work as protective. To be genuinely protective, policy should take into account what is lost by restrictive strategies.

The inadequacy of focus on a particular activity becomes more evident in the paper by **Kirrily Pells**, which uses data from Young Lives surveys in four country sites to show how risk relates to poverty. Particular children suffer from multiple risks. To protect them, intervention must attend to the political-economic context rather than to particular areas of risk in isolation.

Interventions

Is there any way forwards? The papers in the first section showed how protection can go wrong by failing to take account of political and contextual issues, failing to listen to communities and the children on how they see the problems in their lives, and consequently failing to understand what kind of protection might be suitable. The papers in this section show how better understanding is obtained by attending to context and listening to children.

Philip Cook and **Cheryl Heykoop** describe an action-oriented study of the use and dangers of information technology for children in Thailand. While this technology can expose children to abuse, it also provides a resource that they use for their own protection. This study shows the practical importance of consulting the children on how they use the technology, and how their views can enhance its protective potential while dangers of it being used for abuse are minimised. It also shows the possibility of effective community-based intervention.

Natalia Streuli considers Juntos, a social protection programme aimed to help vulnerable children in Peru; it provides cash payments to families in need on condition that they attend to the health and education of their children. She shows how the programme has benefited poor families and their children in a variety of ways. However this does not protect certain children from violence and exclusion, some of which is a consequence of their attendance at schools and clinics. Streuli suggests that such programmes need to be incorporated into broader strategies for children.

These two papers show the advantages of basing intervention on detailed knowledge of a community, and the importance of such knowledge in monitoring the effects of intervention. Even when sound research is carried out, however, there is no guarantee that intervention agencies will get things right for the children (see Woodhead 2007). In the final case study, **Martha Nelems** and **Vanessa Currie** describe research with children that had some immediate impact on how intervention was conducted. Contact with the children led to a better understanding of the problems they faced, better prioritisation in terms of their needs, and some benefits for all children (such as reducing violence in schools). A highly politicised environment, however, and bureaucratic constraints of large organisations, meant that some organisations were not able to react quickly to all recommendations arising from the research. Moreover, many of the lessons were soon lost as staff changed in the organisations.

A final paper returns to the questions raised in the introduction and reflects on what might be learnt from the studies presented here.

A note on research ethics

Ethics of research is particularly problematic when children are involved, partly because power relations between children and adults make it difficult to ensure that children consent to their participation in the research, and partly because their limited experience may make it difficult to understand what they are consenting to and the possible implications for them of the research. There is not space in this issue, still less in each of the papers, for discussion of complexities of research ethics that arise. We as editors, however, believe that all the articles that appear arise from research conducted according to high ethical standards, based on concern and respect for those participating in the research.

Notes

1. Young Lives is core-funded from 2001 to 2017 by UK aid from DFID, and co-funded by the Netherlands Ministry of Foreign Affairs from 2010 to 2014. Sub-studies are funded by the Bernard van Leer Foundation and the Oak Foundation. Further information is provided in the Preface to this collection.

2. The fifth draft of the 'Child Protection Model Law' is available at www.protectionproject.org/wp-content/uploads/2010/11/CP-Model-Law_5th-Draft_Sep-2011.pdf (accessed 19 April 2012). The cited wording has been modified in the 6th draft, in circulation at the time of finalising this article, but not yet in the public domain.

References

Bourdillon, M., D. Levison, W. Myers, and B. White (2010) *Rights and Wrongs of Children's Work*, New Brunswick, Canada: Rutgers University Press.

Crivello, G., U. Vennam, and A. Komanduri (2011) '"Ridiculed for not having anything": children's views on poverty and inequality in rural India', in J. Boyden and M. Bourdillon (eds.) *Childhood Poverty: Multidisciplinary Approaches*, Basingstoke, UK: Palgrave Macmillan, pp. 218–40.

Dercon, S. (2012) 'Understanding child poverty in developing countries: measurement and analysis', in J. Boyden and M. Bourdillon (eds.) *Childhood Poverty: Multidisciplinary Approaches*, Basingstoke, UK: Palgrave Macmillan, pp. 52–72.

Dodge, K. A. (2011) 'Context matters in child and family policy', *Child Development* 82 (1): 433–42.

Ferguson, H. (2011) *Child Protection in Practice*, Basingstoke, UK: Palgrave Macmillan.

Hart, J. and C. Lo Forte (2010) 'Protecting Palestinian Children from Political Violence: The Role of the International Community', Refugee Studies Centre Policy Briefing, Oxford, UK: Refugee Studies Centre 44.

Lonne, B., N. Parton, J. Thomson, and M. Harries (2009) *Reforming Child Protection*, London, UK & New York, USA: Routledge.

Pinker, S. (2011) *The Better Angels of Our Nature: Why Violence Has Declined*, New York, USA: Viking.

Wilkinson, R. and K. Picket (2009) *The Spirit Level: Why More Equal Societies Almost Always Do Better*, Harmondsworth, UK: Allen Lane.

Woodhead, M. (2007) 'Developing policies on child labour: has research made a difference?', in K. Engwall and I. Soderlind (eds.) *Children's Work in Everyday Life*, Stockholm, Sweden: Institute for Futures Studies, pp. 15–24.

Beyond war: 'suffering' among displaced Congolese children in Dar es Salaam

Gillian Mann

This paper examines Congolese children's experiences of war and displacement in the context of the material, social, and relational aspects of their lives in Dar es Salaam. It argues that the challenges, privations and indignities of daily life in urban Tanzania were characterised by feelings of loss, deprivation and hardship so intense they were felt by many to be as or more devastating in their brutality than was life in the midst of war. In so doing, the paper raises profound questions about what kind of 'protection' we are providing to children in these circumstances, and for what purpose.

Au-delà de la guerre : la «souffrance» parmi les enfants congolais déplacées de Dar es Salaam
Cet article examine les expériences des enfants congolais relatives à la guerre et au déplacement dans le contexte des aspects matériels, sociaux et relationnels de leur vie à Dar es Salaam. Il soutient que les défis, privations et indignités de la vie quotidienne en Tanzanie urbaine se caractérisaient par des sentiments de perte, de privation et de difficultés si intenses qu'ils étaient ressentis par beaucoup comme aussi ou plus dévastateurs dans leur brutalité que la vie au milieu de la guerre. Ce faisant, l'article soulève de profondes questions sur la sorte de « protection » que nous offrons aux enfants dans ces circonstances, et à quelle fin.

Além da guerra: "sofrimento" entre as crianças congolesas desalojadas em Dar es Salaam
Este artigo examina as experiências das crianças congolesas frente à guerra e ao desalojamento no contexto dos aspectos materiais, sociais e de relacionamento de suas vidas em Dar es Salaam. Argumenta-se que os desafios, privações e a indignidade da vida cotidiana na zona urbana da Tanzânia eram caracterizados por sentimento de perda, privação e dificuldades tão intensas que foram sentidos por muitos como tão ou mais devastadores em sua brutalidade do que a vida no meio da própria guerra. Ao fazer isto, o artigo levanta questões profundas sobre qual a finalidade e que tipo de "proteção" estamos oferecendo às crianças nessas circunstâncias.

Más allá de la guerra: el 'sufrimiento' de los niños y niñas congoleños desplazados en Dar es Salaam
Este ensayo analiza las experiencias de la guerra y el desplazamiento de niños y niñas congoleños en Dar es Salaam, en los aspectos material, social y relacional de sus vidas. El ensayo revela que muchos niños y niñas describieron los desafíos, las privaciones y la indignidad de sus vidas cotidianas en las zonas urbanas de Tanzania con expresiones de pérdida, privación y dificultad

aún más dramáticas que la brutalidad que experimentaron durante la guerra. El ensayo plantea interrogantes sobre el tipo de "protección" que se está ofreciendo a los niños y niñas en estas circunstancias y para qué.

Introduction

Today there are astonishing numbers of children around the world who have been forced to flee war, violence, and persecution. Millions have crossed into neighbouring states where they live officially and unofficially as refugees. Multilateral and non-governmental agencies have long been concerned with the special protection needs of boys and girls in these circumstances, for whom the social and psychological threats posed by war are considered to be so great as to warrant significant intervention and assistance. Family reunification, rehabilitation, long-term psychosocial therapy, health care, schooling, and vocational training have been provided to large numbers of refugee children in different settings. The intention of these programmes has been to help boys and girls to resume their everyday routines, to seek and give support to their peers and families, to come to terms with their war-related experiences, and, ultimately, to feel out of harm's way, safe from the terrors, disorientation, and violence of war.

This attention to the practical and emotional needs of war-affected children is undoubtedly important but the assumptions that underlie these efforts require careful examination. Not all children experience and make sense of war in the same way. Past experiences of violence may not be the only, or indeed even the primary source of stress and suffering in their lives. Through an examination of the experiences and perspectives of Congolese boys and girls in Dar es Salaam, Tanzania, this paper elucidates the everyday preoccupations of one group of displaced children. Based on 27 months of professional and doctoral research (2001–02; 2006; 2007–08) with more than 100 children between the ages of seven and 18, it aims to break the apparent inseparability between the concepts of being 'war-affected' and being a 'refugee'. Ultimately, its purpose is to show that in order to improve the protection of displaced boys and girls, we need to focus not only on why children became refugees but also on what it is like to *be* a refugee.

The longer version of this article on the Young Lives website situates the study within an active and growing body of research and in so doing, provides a more complete discussion of theoretical perspectives and a longer list of references.

Didier

One day in late 2007, Didier, 18, told me a story that highlighted in a very succinct way the significance of this distinction. In the year or so that we had known each other, I had learnt a lot about Didier. I knew that he had come to Dar es Salaam in 2004, from his hometown of Goma, in North Kivu. I also knew that his move to Dar was not the first time that he had been forced to flee violence and civil conflict: he had also done so in 1994, 1996, 1998, and 2003, returning in each instance after a period in a neighbouring region or country.[1] I had heard from him, his parents, and his five siblings, about the two years that he and his sister, Marie-Claire, now 17, had lived alone in Goma before they were eventually reunited with their family in Dar. While these and other past experiences featured in our discussions, for the most part they were offered as background information to the issues and concerns that

Didier wanted to talk about with me. Like nearly all of the older children whom I knew, Didier's focus was not on the past but on the 'miserable' character of his present life and the challenges it posed to his dreams and expectations for the future. He was certainly willing to talk about the war and his life in DRC, but he repeatedly told me that life in Dar was far worse than anything he had ever experienced in Congo.

It was with comments such as these in mind that I listened attentively to Didier's stories of growing up. The imprisonment of his father, the panicked departures from Goma to seek safety in South Kivu or Rwanda, the constant opening and closing of his school, the death of his young sister, and his struggles to provide for himself and Marie-Claire when the rest of the family went missing – these experiences he described in detail, with one leading into the next. At one point, he mentioned that, at age 14, he had been abducted by the *Forces Démocratiques de Libération du Rwanda* (FDLR) to fight as a child soldier (Hubert 2011). While travelling to visit his uncle, his bus was stopped in the middle of a national park and he and several other boys were forcibly taken by a group of heavily armed men to a high-walled, empty building in the middle of the forest. Here Didier stayed for a number of weeks, during which time he and others were forced to clear the forest and to build another building. Having eaten very little food, most were weak with hunger. Those who failed to work effectively were summarily killed. One day, he and many other boys were pushed inside another lorry and told that they were going to fight. As the vehicle barrelled along the road, Didier and four others made the decision to jump off: '*I knew that either way I was going to die, and this way seemed better*', he told me. Together, the boys chose their moment and threw themselves from the speeding lorry into the bush beside the road. The vehicle came to a screeching halt and the driver and a soldier jumped down and began shooting at them. But these men did not know how many boys had jumped, and when they saw the open skulls of two of the escapees, they stopped shooting, announcing that '*everyone has died*'. When the lorry had driven away, Didier, suffering from an injured shoulder, waited for a while before stumbling into the bush and beginning a two-day walk to the first village he found. Initially, the other surviving boy accompanied him but eventually he succumbed to the bullet in his leg and ceased to travel with Didier. After a few days in the village, Didier was given a lift with a local driver back to Goma, where he was reunited with Marie-Claire.

For some reason, hearing this particular story, alongside all of the others, I felt quite shocked. My shock was not out of ignorance of the abduction, forced labour, militarisation, and brutal killings of children that have taken place throughout the past 15 years of civil conflict in Congo. I have known many boys and girls who have experienced life in these circumstances. My amazement came from the countless times that Didier and others had asserted to me how much better life had been in DRC compared to their 'miserable' existence in Dar es Salaam. '*Life there was difficult*', he told me, '*but at least we weren't refugees*'. Perhaps my shock was caused by hearing about these horrific experiences while sitting with a relatively healthy 18-year-old, drinking Coke under a shady tree, in what, to all objective observers, was a peaceful country like Tanzania. Suddenly the incongruous nature of it all overwhelmed me. How is it, I wondered, that a person can have experienced such horrific things as Didier described, and still say that life in that context was preferable to life away from the war? What is it about life as a refugee in urban Tanzania that is so bad that it led so many people – young and old – to describe life in Congo as better?

'Suffering' among refugees

In Dar es Salaam, the feeling that life was 'wretched' (Swahili: *unyanyapaa*) was common for Congolese of all ages and in a variety of circumstances. Their undocumented status and the hostile environment in which they lived meant that most adults and children lived in constant fear of exposure and its dreaded consequences. Most were highly distrustful of Tanzanians, Tanzanian authorities, international agencies and, often, other Congolese. Their illegal status meant that they needed to exercise extreme caution in their efforts to find paid work, access services or assistance from churches, government, or NGO sources, and in their interactions with unknown individuals who might be able to assist them with employment, housing and other ways of addressing their material needs. These challenges were experienced alongside sometimes tense and difficult relationships between parents and children. The result was a life characterised not only by extreme poverty and a lack of basic social services such as health care and education, but also by high levels of fear, isolation and disappointment, and low levels of social support.

For many boys and girls, it was not so much the inability to attend school, the lack of decent shelter or the need to hide their identity, per se, that were so difficult. It was the way living in these 'difficult surroundings' (*mazingira magumu*) made them feel. Most described an overwhelming sense of abjection, a term Ferguson (1999: 236) uses in relation to the experience of long-standing decline on the Zambian copper belt, to describe the sense of not only losing one's place in the world, but the associated feelings of betrayal, humiliation, and having been cast aside. These feelings were exacerbated by a sense of being suspended in a boring and meaningless present in which the future was so uncertain and beyond control that it was at times unimaginable. Boys' and girls' day-to-day lives and futures were contingent on so many different things and people that many found it, both literally and figuratively, painful to think beyond the here and now.

So overwhelming were these bad feelings that many, like Didier, said that they would rather live in DRC, even in the midst of war, than in Dar es Salaam. Given the profound insecurity and overwhelming challenges posed by life in DRC, it is significant that so many children and young people made similar statements. The fact is that life in eastern DRC has been extraordinarily difficult for generations. State-sponsored violence, terror, and pillage began in earnest under King Léopold and the Belgian colonial authorities and continued under the long dictatorship of Mobutu Sésé Seko (Prunier 2009). Since the mid-1990s, the provinces of North and South Kivu, Ituri, and Maniema have been engulfed in civil war and cross-border conflicts. So great is the scale of the crisis that in 2007 the International Rescue Committee estimated that 5.4 million people had died since 1998, in large part from curable illnesses such as malaria, diarrhoea, pneumonia, and malnutrition (Coughlan *et al.* 2007). Extreme brutality, massive population displacement, widespread sexual violence, and the collapse of public services such as health, education, and infrastructure have characterised this period. Every person I have known from eastern Congo – adults and children, rich and poor, rural and urban, refugees and those who remained – has been touched by this large-scale humanitarian crisis. While numerous peace agreements have been signed, and the war officially ended in 2003, political violence and civil conflict continue in the east. Reports of widespread killing, rape, and other atrocities on the part of government troops and rebel groups are still commonplace.[2] Many aid agencies are warning of impending large-scale humanitarian disaster.[3]

Despite all of these problems, many children nonetheless reflected on their lives in Congo with a sense that perhaps it was not so bad after all. It was violent and unpredictable. Insecurity had forced them to flee. '*But there we were free, here we are nothing*', Hamisi told me. When I asked him what he meant by freedom, he explained that as a refugee, '*You are no longer a person ... You have lost not only your land, your things and your country, you have lost yourself*'. '*In Dar es Salaam*', Marie-Claire said, '*we are no longer human beings*'. Alain echoed:

> *Here we live like chickens, caged up ... circling around, never knowing if we will be released. In Congo, it was different, I had a life. I was a person.*

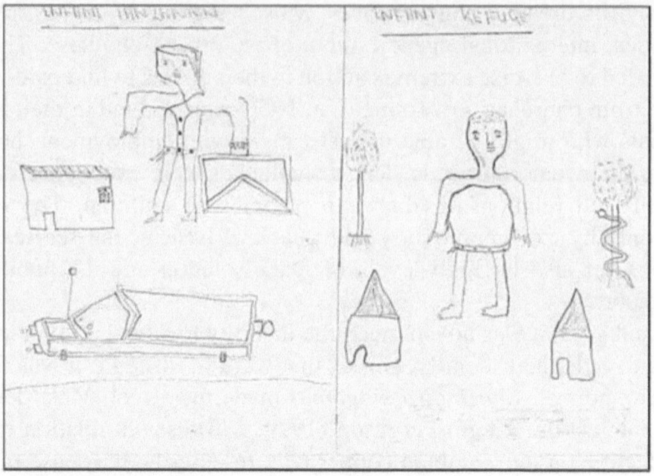

Figure 1: Drawing by Christophe, 14

This idea of 'freedom', and of 'having a life', defined residually by reference to the absence of freedom, and the absence of 'a life', were central to children's understanding of their suffering in Dar es Salaam. These feelings may have been particularly acute among the children I knew, who tended to come from families with relatively high levels of formal education, who had primarily urban and professional backgrounds. Despite – or perhaps because of – the disappointments and failures associated with their lives in Dar, their parents often spoke to me of a desire to instil in their sons and daughters a sense that life could be different, that despite their hardship, they were nonetheless people who could make a good future for themselves. Some, like Mireille, mother of six, asked me to spend time with her 17-year-old daughter, to talk with her about my life, '*to show [her] another world*'. Others, like Agnès, encouraged me to take adolescent Lucie to the University of Dar es Salaam so that she could '*see a possible life*'. Despite these attitudes, in the face of formidable barriers to their achievement, such strong hopes often faded and disappeared. What remained

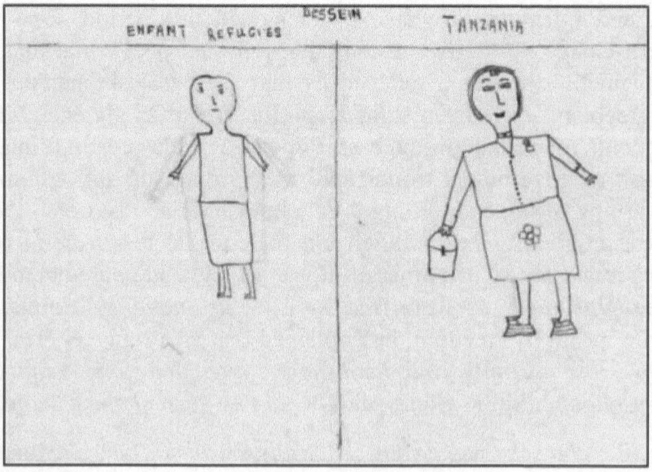

Figure 2: Drawing by Angela, 12

Figure 3: Drawing by Luc, 13

was a sense of failure, abjection and sometimes, hopelessness: '*We have fallen*', 18-year-old Martin told me, '*and we don't know how to get up*'.

Suffering and refugeeness

It was this sense of abjection, and its impact on individuals' sense of agency that children were referring to when they said '*tunaishi kwa mateso*' (we are suffering, or we are living with suffering). Not only did boys and girls have the sense that they had been rejected and discarded by external forces, but they struggled against internalising the idea that they were, in fact, the useless and lazy beggars, the 'garbage' on the street, that they believed Tanzanians thought them to be. One of the many ways that I tried to elicit children's perspectives on these issues was to ask them to draw a picture of two children, one Tanzanian, the other a refugee. The following three drawings (Figures 1, 2, 3) represent a few examples of the many that I collected according to these instructions.

On their own, these drawings give a general indication of the way Congolese children saw themselves in relation to Tanzanians: smaller, poorer, less healthy, with fewer belongings and fewer resources. But it was in discussing these drawings that I was able to learn more from individual children about how precisely they saw themselves, and refugees in general, in relation to the world around them. Asked what the two people were thinking in Figure 1, 14-year-old Chistophe replied that the Tanzanian child was thinking what kind of car he would one day own, while the refugee child was tired, hungry, and sad. I asked why. '*Because he has nothing else to think about*', said Christophe. Similarly, in Figure 2, 12-year-old Angela explained to me that the person on the right, a Tanzanian girl, was healthy and strong, had nice clothes, shoes, hair extensions, a purse, and jewellery. The refugee child on the left, she said, had a simple, plain cloth for a dress, no shoes, no hair, no accessories. When I asked her to tell me why there were such big differences in the way these two girls looked, she said it was because Tanzanians are '*more intelligent*'. They know how to get money and live well and God looks after them (note the cross on the child's necklace). The refugee child, she said, does not know how to get these things. Luc, 13, made similar remarks about the two children he drew in Figure 3, emphasising not only how the refugee child was small and undernourished, with torn clothes and skin lesions, but how his situation contrasted with that of the confident, well-dressed Tanzanian child, who stood at the forefront, unafraid to be seen or noticed.

These children's feelings that, as refugees, they were somehow not able to meet their full potential were also shared by adults. In the planning session for a half-day-long workshop for seven- to 11-year-olds in April 2002, my two Congolese female co-facilitators requested that I not include one of the activities I had prepared, a task I had called '*I am strong*'. This simple activity involved each child drawing the outline of one of their hands on a blank piece of paper. The idea was for each boy or girl to identify five things – one for each finger – about themselves or their families that they felt good about. Both women argued that the activity would be disempowering for children because most felt that there was nothing good to say about themselves. This view, that the failures that accompanied refugee life – and the failure inherent in being a refugee (see Harrell-Bond 1986: 6) – were somehow embodied in children, was shared by other adults as well. Six years later, again in Tanzania, Ismael told me in front of his 16-year-old son, James, and two of James' friends, that Congolese children in Dar were '*a lost generation*': according to Ismael, these boys and girls had realised that they had no future to speak of, and had begun to behave accordingly. '*As refugees, they will never become anything here in Dar es Salaam*', he asserted. This association between being a refugee, being a child, and having no future was central to the way in which 'suffering' was understood.

In fact, for the Congolese children and adults I knew in Dar es Salaam, the concepts of suffering and refugeeness were inseparable. When one suffered, one became a refugee (*mkimbizi*): those who saw themselves as having no choice, no future, and no control over their own lives, considered themselves and were considered by other Congolese, to be refugees (*wakimbizi).* So linked were suffering and refugeeness that they were often evoked together: '*To be a refugee is to suffer*', 16-year-old Mugisho told me, after explaining why his father, a trained physician from Bukavu, was unable to find paid work. '*Me, as a refugee, I suffer*', said adolescent Lucie, when lamenting her mother's requirement that she remain at home all day. '*The life of a refugee is a life of suffering*', said Charles, after he withdrew from school and moved house because of his inability to pay the 'hush money' the school principal said he required in order to not report Charles' family to the police. The circumstances that elicited these sentiments were not in all cases unique to refugees; there were other marginalised groups in Dar es Salaam who had similar experiences. The social and economic conditions under which most undocumented Congolese lived were in many ways not particular to them. But the feelings of abjection, of futureless-ness, of the inability to escape the crushing power of the present – these were the things that made these children feel like refugees.

Thinking too much

Boys and girls struggled not to feel overwhelmed by their suffering. Doing so required thinking about the future – about the possibility of a different, better life – despite the difficulty and painfulness of doing so. It necessitated a delicate balance between thinking, and not 'thinking too much' (*kichwa na kinajaa na mawazo,* literally 'my head is full of thoughts'). Children like Didier needed to think about what the possible options were for him to survive and make a life for himself. Was going back to school a possibility? What kind of job could he expect to find? Was it best to remain in Dar es Salaam, to return to DRC, or to migrate elsewhere? Who did he know who could help him? Thinking through these questions was critical to his ability to imagine a better life for himself. At the same time, it was a challenging cognitive and practical feat to think of the future without also focusing on the present. Didier's future plans, and those of his peers, depended in large measure on their present conditions. Yet Congolese of all ages argued that the best way to cope with the hardships and uncertainties of life was to avoid thinking too much about them because doing so could lead one to fall ill with headaches, 'pains in the heart', stomach pains, and debilitating exhaustion and insomnia. Even more worrying was the view that 'thinking too much' could result in feelings of hopelessness and despondency, and, at its worst, madness. The gravity of these physical and emotional symptoms contributed to a widespread fear among both adults and children of falling ill with 'too many thoughts' (*mawazo mengi*).

Thoughts of violence and war are often assumed by researchers and practitioners to be the cause of problems like 'thinking too much' among refugee children. Yet among those I knew in Dar, it was the severe hardships of daily life and the challenges these posed to individuals' ability to plan for the future that were said to lead to 'too much thinking'. Thierry told me that, with no money, no job, and 'nothing to do', he had too much time to think about his life, his lack of prospects, his inability to support his family back home in DRC, and the frightening reality that he might live the rest of his days in exactly these circumstances. When she was out of school and spending her days confined with her sister to her small, one-room house, Lucie said she began to 'have thoughts' (*kuwa na mawazo*). Her mother's inability to find paid work, her own stalled academic progress, the cruelty of her neighbours, her hungry belly, and her lack of ability to imagine a means of escape led her to feel overwhelmed, exhausted and, at times, hopeless about the future. Adults also spoke to me of how the brutal present and the unknown, unimaginable future sometimes led them to fall ill with 'too many thoughts'.

Small boys and girls were also said to suffer at times from the problems associated with 'too much thinking'. While very young children were not understood to worry about the future, per se, their ailments were often understood by adults and older children to be manifestations of the stresses of life in Dar. During the time that I knew her, three-year-old Lise frequently complained to her parents that her 'head [was] sick'. 'Pains in the head' were commonly reported by young children and, in this case, Lise's parents said that she was sick because she was confined to her house, she had no other children to play with besides her 18-month-old brother, and she had a significantly diminished appetite. Lise also cried a good deal and was often listless – two additional signs that she was 'sick because of this life'.

The importance of 'thinking too much' as an idiom of distress among Congolese in Dar es Salaam lay in the meanings adults and children attributed to the condition, which was considered extremely serious by everyone I knew. One interpretation of this type of somatisation, offered by some anthropologists and post-colonial theorists, may be relevant here. These authors suggest that conditions like 'thinking too much' are the embodiment of extreme social and structural inequalities, in which the bodies of the powerless become a means of expressing opposition to their powerlessness (e.g., Das 1992; Diouf 2005; Weiss 2005). To be sure, those who suffered from 'thinking too much' in Dar es Salaam were those who were very poor, marginalised, isolated, and who felt an overall sense of abjection, helplessness, and hopelessness about the future. Maire-Claire told me:

> *Live, die. It is the same thing. No one cares anyway.*

'Thinking too much' may have been a means for people to both express these feelings of despondency and to protest against them (see Eyber 2001: 134). The worse their social and economic situation became, the more difficult it was to escape the pain and difficulties associated with 'thinking too much'. In this way, their feelings of subjectification and their protests against it may have been inscribed on their bodies (Mbembe 2001). As a mode of social consciousness, 'thinking too much' was both individual and collective in that the sources of the condition were understood to be shared with others in similar circumstances.

Beyond war

The Congolese children I knew in Dar es Salaam were engaged in an ongoing struggle to avoid becoming overwhelmed by and trapped in a brutal, relentless, and never-ending present. Their fear and dread of 'thinking too much' was related not only to the emotional, existential, and physical pain associated with the condition, but also to the truth that lay behind it. This truth was that the social and economic conditions of displacement could succeed in undermining their very sense of who they were. When the possibility of becoming was unimaginable and non-existent, their suffering threatened to undermine their basic humanity because it was the very idea of a future, even if as yet unrealised, that motivated them to make it through the day ahead. One endured the present in order to escape it in the future. This way of looking at life suggests that for some children, the future, or at least the imagining of the future, defines the present. Without a future, and without the ability to think about, plan for, or imagine a future, their present seems meaningless and their existence little more than that of 'animals'.

On a theoretical and conceptual level, this understanding of the source of refugee children's suffering challenges at least one dominant assumption in social scientific research with refugees, namely that the experience of war-related violence fundamentally and irrevocably alters the way in which people see themselves and their relationships with others (e.g., Nordstrom 1997). This presumption that violence exerts an all-pervasive, overwhelming influence on all aspects of personal, social, economic, and cultural life has been the silent frame for the main question that has been asked about refugees and others who have lived in conflict zones, namely: 'How are people getting on with life now that their whole worlds have been destroyed by violence?'. Behind this and other, related questions, has been the view that the violence of war is somehow extraordinary, a profoundly shocking experience that takes place outside the 'normal' range of human experience. Coping with war-related violence has thus been understood to be of critical importance to refugees, and therefore to research about refugees (Lubkemann 2008: 10).

This assumption has had particular purchase in research with displaced children in Africa. So widespread is the assumption that war exerts a deleterious effect on all aspects of child development that the lives of refugee boys and girls in these contexts have rarely been explored through any other lens. The abundance of reports and studies about former child soldiers, girls who have served as sex slaves, and those who have been forced to kill, torture, and rape their parents are but a few examples of this trend. Although these works raise many significant issues and concerns, they only give part of the picture. They draw our attention away from the mundane, everyday lives of children – their routines and activities, their social interactions and relationships, the character of their work and play, and the things that they think about on a daily basis. From the perspective of refugee boys and girls in Dar es Salaam, these ordinary aspects of daily life were as much, or more, a part of who they were as the war they left behind in DRC.

My research is not the first to challenge this assumption that the experience of war-related violence fundamentally alters cultural modes of being and interacting in all settings. A growing body of ethnographic research with refugees and displaced people in numerous contexts has shown that the experience of violence is mediated differently by different people, at different times, in different settings. These works demonstrate the integral role played by society and culture in shaping the meaning people attribute to violence and the role of context in determining the subjective interpretation of human experience more generally. For example, in his research with former female combatants in Mozambique, West (2000) found that commitment to the ideology of FRELIMO powerfully influenced the way in which young women fighters experienced the war because they felt that they were making a valuable contribution to the creation of a more just and equal society. These women were able to make sense of the violence that they experienced and perpetrated because they saw a meaningful reason behind it. Interestingly, West (2000: 191) argues that in the aftermath of the war, it was more difficult for these young women to come to terms with the failure of these larger dreams of liberation than it was to cope with the violence of war. He and other authors assert that the experience of violence cannot be understood without reference to the social, cultural, and moral contexts in which people live (Boyden and Mann 2005: 17).

Despite the overall salience of these assertions, they have been applied in far greater measure to our understanding of the experiences of refugee and war-affected adults than they have to children. In part, this has to do with the reality that, until recently, the vast majority of what we knew about children and war we learnt from developmental psychologists working with children in North America and Europe. While some studies have been undertaken in Africa, these have generally relied on generic behaviour checklists and assessments made in the course of therapeutic interventions for individual children (e.g., Garbarino and Kostelny 1996; Morgos 2009). Because continuity, stability, predictability, and physical and emotional security are considered fundamental to children's well-being and development, exposure to violence is assumed to have a direct and automatically damaging effect. Consequently, refugee children are viewed in this literature largely as traumatised victims who need medical and psychological care in order to resume 'normal' lives. Furthermore, as Boyden and de Berry (2004) have argued, their responses to violence are understood to be more or less the same in all contexts. The term 'suffering' is often used by both psychologists and anthropologists to refer to the serious wounds to the body and spirit that war and displacement can cause. The cause of refugee and war-affected children's 'suffering' is usually assumed to be the experience of war-related violence.

The findings of my research with undocumented Congolese children in Dar challenge this assumption. For these boys and girls, daily life in urban Tanzania was considered more threatening to their individual well-being than was their experience of conflict and war-related violence in DRC. As 16-year-old Munga told me:

> *When I can't sleep at night it is not because I have bad dreams about the war. It is because I don't know what I am going to do tomorrow or the next day or the next day. How will I ever leave this miserable life?*

I am aware of the dangers of taking such statements at face value. Sometimes the present can seem worse than the past because of its proximity. Some children may experience serious and prolonged psychological and emotional distress as a result of the violence they experienced (and sometimes perpetrated) in conflict zones. However, the analytical and methodological problem is that researchers who look for traumatic responses to experiences identified in advance as highly stressful may miss important differences in the individual and collective experiences of boys and girls in different contexts (Boyden and Mann 2005: 13). Indeed, at the stage of their lives during which I knew them, the focus of Congolese children was not on the past but on transcending the crushing power of the present – the bad feelings and debilitating thoughts that resulted from living in a xenophobic, discriminatory environment in which they lacked basic entitlements – in order to plan for or imagine a future.

Conclusion

Congolese children's lives in Dar es Salaam were influenced by a number of social, economic, political, and personal factors that had little to do with the war they had experienced and fled. The challenges, privations, and indignities of daily life in urban Tanzania were characterised by feelings of loss, deprivation, and hardship so intense that they shaped to a large extent the ways boys and girls came to think about and tried to make sense of their lives. Existential concerns about things like 'freedom', 'having a life', and 'feeling like a person' were at the heart of their condition. Displacement had affected their lives in countless ways. But from their perspective, its greatest impact was on their sense of themselves as *people*. The social and economic conditions in which they lived led them to question not only who they were and their place in the world but whether they were in fact human beings. They queried not only their own 'suffering' and those of their families; they questioned life, its meaning, and the existence of God. Living with, making sense of and challenging their existential predicament was a central feature of their daily lives.

These findings raise profound and challenging questions for those of us concerned with child protection. What kind of 'protection' are we actually providing to children like Didier and Munga and the many others who live in similar circumstances, and for what purpose? It is true that in Dar, these boys and girls were safe from the violence of war. They no longer feared that their families would be killed, that their school would be attacked, or that they would be forcibly made to fight. They were 'protected' from these grave risks, but at the same time made intensely vulnerable to violence of a different and more insidious kind. Fear of physical death at the hands of soldiers and rebels was replaced by fear of a spiritual death, in which life had no meaning. War did not kill them, but they feared that life as a refugee might. In this context, protection is about more than safety.

Those of us concerned with the protection and well-being of children need to engage with these realities in substantive and meaningful ways. On a theoretical level, doing so requires acknowledging the psychic damage that hostility, indifference, and extreme poverty can cause. It means rethinking the important role that context plays in shaping personal and collective responses to adversity and doing away with simplistic assumptions about which experiences and aspects of life are most detrimental to boys and girls. On a practical level, it means creating the conditions under which children can develop and sustain their hopes and aspirations, even in the most difficult of circumstances. Doing so requires seeing boys and girls not only as *children* who need protection from physical harm, but as *human beings* who seek and deserve to live lives worth living.

Notes

1. The eruption of Nyiragongo volcano in Goma in January 2002 also forced Didier and his family to flee for several weeks to safety in Rwanda.
2. Some of these agencies include Human Rights Watch (www.hrw.org/en/news/2009/12/12/un-act-end-atrocities-eastern-congo); the International Rescue Committee (www.theirc.org/special-report/congo-forgotten-crisis.html); Refugees International (www.reliefweb.int/rw/rwb.nsf/db900sid/MUMA-82U6NL?OpenDocument&rc=1&cc=cod); and the United Nations Office for the Coordination of Humanitarian Affairs (www.reliefweb.int/rw/RWFiles2010.nsf/FilesByRWDocUnidFilename/FBUO-82UK4T-rapport_complet.pdf/$File/rapport_complet.pdf).
3. The FDLR is the name given to groups of rebels composed of some Rwandan Hutu refugees who fled to DRC in the wake of the 1994 genocide in Rwanda. Violence instigated by and inflicted against this group has been continuous over the past 15 years or so, as regional politics have played themselves out in complex and brutal ways on the people of eastern Congo.

References

Boyden, J. and J. de Berry (2004) 'Introduction', in J. Boyden and J. de Berry (eds.)', *Children and Youth on the Front Line: Ethnography, Armed Conflict and Displacement*, Oxford, UK: Berghahn.

Boyden, J. and G. Mann (2005) 'Children's risk, resilience and coping in extreme situations', in M. Ungar (ed.) *Handbook for Working with Children and Youth: Pathways to Resilience Across Cultures and Contexts*, Newbury Park, CA: Sage.

Coughlan, B. *et al.* (2007) 'Mortality in the Democratic Republic of Congo: An Ongoing Crisis', New York, USA: International Rescue Committee, available at www.theirc.org/sites/default/files/resource-file/2006-7_congoMortalitySurvey.pdf (retrieved 21 February 2010).

Das, V. (1992) *Mirrors of Violence: Communities, Riots and Survivors in South Asia*, Oxford, UK: Oxford University Press.

Diouf, M. (2005) 'Afterword', in A. Honwana and F. De Boeck (eds.) *Makers and Breakers: Children and Youth in Postcolonial Africa*, Oxford, UK: James Currey.

Eyber, C. (2001) 'Alleviating Psychosocial Suffering: An Analysis of Approaches to Coping with War-Related Distress in Angola', Edinburgh, UK: Queen Margaret University College, unpublished PhD dissertation.

Ferguson, J. (1999) *Expectations of Modernity: Myths and Meanings of Urban Life on the Zambian Copperbelt*, Berkeley, CA: University of California Press.

Garbarino, J. and K. Kostelny (1996) 'What do we need to know to understand children in war and community violence?', in R. Apfel and B. Simon (eds.) *Minefields in Their Hearts: The Mental Health of Children in War and Community Violence*, New Haven, CT: Yale University Press.

Harrell-Bond, B. (1986) *Imposing Aid: Emergency Assistance to Refugees*, Oxford, UK: Oxford University Press.

Hubert, T. (2011) 'Will trials bring peace to DR Congo's Ituri region?' BBC News – Africa, 27 April 2011, available at www.bbc.co.uk/news/world-africa-13199899 (retrieved 29 February 2012).

Lubkemann, S. (2008) *Culture in Chaos: An Anthropology of the Social Condition in War*, Chicago, IL: University of Chicago Press.

Mbembe, A. (2001) *On the Postcolony*, Berkeley, CA: University of California Press.

Morgos, D. (2009) 'Psychological Impact of War Atrocities in Darfur on Children: Implications For Assessment and Interventions in War Zone Areas', paper presented at the Children and War Conference at Rutgers University, Camden NJ, USA, 3–5 April, 2009.

Nordstrom, C. (1997) *A Different Kind of War Story*, Philadelphia, PA: University of Pennsylvania Press.

Prunier, G. (2009) *Africa's World War: Congo, the Rwandan Genocide and the Making of a Continental Catastrophe*, Oxford, UK: Oxford University Press.

Weiss, B. (2005) 'The barber in pain: consciousness, affliction and alterity in Urban East Africa', in A. Honwana and F. De Boeck (eds.) *Makers and Breakers: Children and Youth in Post-Colonial Africa*, Oxford, UK: James Currey.

West, H. (2000) 'Girls with guns: narrating the experience of war of Frelimo's "female detachment"', *Anthropological Quarterly* 73 (4): 180–94.

Protecting children from trafficking in Benin: in need of politics and participation

Neil Howard

This paper critically examines policy in Benin against child trafficking. Drawing on interviews and participant observation with adolescent labour migrants and their communities, it problematises both the assumptions underpinning anti-trafficking policy and the appropriateness of the initiatives that comprise it. It suggests that, in order truly to protect the young labour migrants defined as trafficked, the policymaking establishment needs to focus more closely on the structural economic underpinnings of migration and exploitation, and to adopt a more participatory approach to policymaking.

Protéger les enfants de la traite au Bénin : la nécessité de politiques publiques et de participation

Cet article examine dans un esprit critique les politiques du Bénin de lutte contre la traite des enfants. Grâce à des entretiens et l'observation des participants avec des travailleurs migrants et leurs communautés respectives, il problématise à la fois les suppositions qui étayent les politiques publiques de lutte contre la traite et le caractère approprié des initiatives que ces politiques englobent. Il suggère qu'afin de vraiment protéger les jeunes travailleurs migrants définis comme victimes de la traite, les instances chargées de formuler les politiques publiques doivent se concentrer davantage sur les bases structurelles économiques de la migration et de l'exploitation et adopter une approche plus participative de la formulation de politiques.

Protegendo crianças do tráfico em Benin: necessidade de políticas e participação

Este artigo examina criticamente a política em Benin contra o tráfico de crianças. Baseando-se em entrevistas e observação participante com imigrantes trabalhadores adolescentes e suas comunidades, o artigo problematiza os pressupostos que apóiam a política anti-tráfico e a adequação das iniciativas que a compõem. O artigo sugere que, para proteger verdadeiramente os jovens imigrantes trabalhadores definidos como vítimas do tráfico, a formulação de políticas precisa concentrar-se mais na base econômica estrutural da imigração e da exploração, e adotar uma abordagem mais participativa na formulação de políticas.

La protección de la niñez ante el tráfico de personas en Benín: la necesidad de políticas públicas y participación

Este ensayo hace un examen crítico de las políticas contra el tráfico de niños y niñas en Benín. Basado en entrevistas a adolescentes que migran en busca de trabajo y en la observación directa en sus comunidades, el ensayo cuestiona tanto los supuestos que subyacen tras las políticas contra el tráfico como si las acciones emprendidas son o no adecuadas. El ensayo demuestra que, para que la protección de los jóvenes trabajadores migrantes en situación de tráfico sea eficaz, los responsables políticos tendrán que atender con más empeño las causas económicas y estructurales de la migración y la explotación y tendrán que impulsar métodos más participativos para la elaboración de políticas.

Introduction

> *People say that moving is bad, but what alternatives do they offer? They have never even been here to find out how we live.* (Interview with T, 30 August 2007)[1]

This paper challenges conventional thinking on an issue that has been identified by the international child protection community as one of the major challenges of the past decade: child trafficking.[2] My involvement in this field dates to 2005, when I began working with a child rights NGO in Cotonou, Benin, and was immediately exposed to institutional and governmental concern over the apparent 'pandemic' engulfing the country. Struck by the disjunction between the lived realities of the young labour migrants with whom I worked and the official representation of them as victims of trafficking, my research has sought to examine the creation, operation, and consequences of this disjunction, with a particular focus on the failures of policy to achieve the protection it claims as its goal. I concentrate here specifically on the way community understandings of adolescent work and migration and the unaddressed structural backdrop of poverty point to major gaps in the discursive and policy framework that can only be addressed through greater community participation in the identification of protection problems and solutions, and a more direct engagement on the part of the policy establishment with the politics of poverty.

Research context and methodological overview

Child trafficking began to emerge as a problem in Benin at the start of the last decade. Though child labour had long been a focus of international and national attention within the country, child trafficking displaced it as the central preoccupation with the development of the 'Trafficking Protocol' and with the advent of the 'Etireno Affair',[3] whereby Benin became tarred as an epicentre of the international traffic in children.

Shamed by this designation, Benin's government quickly ratified the Trafficking Protocol,[4] facilitating an influx of anti-trafficking funds and establishing various anti-trafficking initiatives. As elsewhere in the world, the trafficking of children was defined as any component of the process that coupled movement with exploitation, and exploitation was taken to include any economically productive activity 'unacceptable' for a child and from which a third party derives a profit. As per the International Labour Organisation (ILO)'s anti-child-labour framework, 'unacceptable' activities included work in

> *mining and quarrying; manufacturing; construction; electricity, gas and water; sanitary services; transport, storage and communication; and plantations and other agricultural*

> *undertakings mainly producing for commercial purposes, but excluding family and small-scale holdings producing for local consumption and not regularly employing hired workers.* (ILO 1973: article 5(3)).

My investigations into Benin's anti-trafficking policy have taken me through five countries over 14 months, during which I conducted over 180 interviews with more than 300 people, and observed and worked with individuals and institutions at every level of the policy chain. At the institutional level, I focused on those bodies at the heart of producing and sustaining the anti-trafficking discourse, and at the forefront of forming and implementing anti-trafficking policy, both in Cotonou and throughout Benin.

At the ground or community level – from which most of the data underpinning this paper is drawn – I divided my work into two phases. First, I interned at a shelter for 'rescued children' run by an international NGO, and I spent months working with and interviewing young people identified as victims of trafficking. Second, I chose four case study villages from two districts in the Zou Region, where I interviewed community leaders, community members, and current and former young migrants supposed by the authorities to be victims of trafficking. I chose these villages for two reasons. First, they are from a region that has been institutionally identified as a hub of child labour, trafficking, and exploitation. This is itself a result of the fact that young males frequently migrate to Nigeria for mine work – an activity that has been identified as an archetype of child labour and trafficking. Second, these villages are also at the heart of the Beninese cotton belt, which has seen household incomes plummet and remain low over the last 15 years as a result of international falls in the price of cotton. I therefore wished to explore the hitherto un-examined link between declining income from cotton and adolescent movement to the mines.

In total, I interviewed 170 people across these four villages. I used various participatory techniques in the villages, including principally open-ended individual and group interviews, covering how people see their life-worlds, how they experience and view youth work and movement, how they understand childhood and the transition to adulthood, the links between cotton and movement, and what people have to say about current anti-trafficking interventions. I have consistently discussed my interpretations of interview data with informed colleagues and interviewees.

Dominant narratives and policy prescriptions

Since it is both widely known and has been well documented elsewhere (see, for example, Alber 2011; Hashim and Thorsen 2011; Howard 2011), I will only briefly outline here the dominant narrative around child trafficking and the mainstream policy approach to dealing with it. In Benin, as in many other countries, trafficking is depicted as the main child protection challenge facing the authorities, attaining near pandemic levels. The crime is seen to be comprised of two interlinked components which are themselves individually problematic – child labour and child migration – and is understood to be the result of a number of pathological causes or factors. These factors begin with poverty, which is always treated as a causeless, abstract reality. They also include criminal deviance, poor parenting, and the corruption of traditional mechanisms of social solidarity that involve the circulation of children between families. The general trend in the narrative is that child work and migration are inherently dangerous, that children should be at 'home' with their families and at school for their well-being, and that when they are not, it must be as a consequence of some crisis situation, such as kidnap, a flight from starvation, or parents being tricked into 'selling' their offspring.

Though steps taken by the state and its partners to address trafficking include the small-scale roll-out of micro-credit or the promotion of free schooling, the major emphasis of all efforts has been on preventing child work and the migration that is seen to lead to it. This has involved the passing of a national anti-trafficking law that regulates child movement and makes it legally very difficult, the establishment of village vigilance committees to prevent children migrating for work, the expansion of border surveillance in cooperation with Nigeria, and the massive promotion of sensitisation campaigns aimed at convincing parents that home is where a child should be, as opposed to away and in work. Of crucial importance is that little distinction is practiced between small children and older adolescents who, while legally still minors, are socially closer to adulthood.

The rest of this paper challenges the utility of such approaches by contrasting this dominant narrative with community-level understanding, and by offering a political-economic analysis of the poverty that forms their context.

The view from below

The image that emerges from my research with young migrants and their communities is significantly more complex than that propagated by the institutions promulgating the anti-trafficking discourse. Though, at times, work and migration *can* be experienced negatively by the young, in most cases they are seen as positive realities and constructive life strategies.

Child work

In my case study villages, work, whether inside the home or for remuneration outside it, is not seen as a damaging 'adult' sphere from which under-18s are to be sheltered, nor in fact as anything other than an eminently positive and necessary part of being young and growing up. In most households, children as young as three may be asked to perform basic tasks such as filling pots of water; progressing at five or six to keeping an eye on their very smallest siblings; at eight or nine to washing those younger than themselves or sweeping the courtyard; and at 12 to cooking, cleaning, and taking on the rest of the tasks performed by adult household members at home or within the context of their small-scale economic activities. One village elder in Tenga explained to me the incorporation of young males into the agricultural labour force:

> *Already at eight you can go to the field. At 12 you can begin to work like a man.* (Group 8 Interview, Tenga Village, Zogbodomey, 9 May 2010)

Such realities are not viewed as a grave hardship. For young children, the line between work and play is much blurrier than is perceived by the anti-trafficking policy establishment, and inventive children regularly incorporate pot-filling and sibling care into various entertaining games. Children also regularly profess enjoying the activities they are tasked with, even if at times these activities may be taxing. In one of the many interviews I conducted with groups of adolescents during which we discussed precisely this issue, I asked how they felt about working and whether they liked it, and received in response a cacophony of '*yes*', with the addendum '*because it feeds us!*' (Group 12 Interview, Atomè Village, Zogbodomey, 14 May 2010).

This addendum points to the core of both the positive view of child work and the gradual incorporation of even young children into the working collective. The understanding of work dominant here is integrally related to the need to secure subsistence in what is a materially poor environment. '*We're not in France, after all*', one teenage interviewee declared (Interview with CS, Sehere Village, Zakpota, 12 April 2010), and the young, along with all other able-bodied household members, are thus both needed and expected to contribute to the collective endeavour that is necessary for life. Within this logic, their work, remunerated or otherwise, represents both an immediate necessity and a fruitful way of socialising them into the self-sufficient, responsible adults they will have to become as integral, contributing members to collective well-being.

This is underlined by much widely deployed local-level symbolism. On the one hand, it is 'the thief' who represents the figure of universal revilement in the communities, since the thief embodies the unacceptable principle of taking without giving. On the other hand, I was repeatedly told that young children and adolescents were put to work precisely so that they could have the opportunity to *avoid* taking without giving – doing nothing – and '*so that they can learn how to take care of themselves*'. Crucially, in this context, 'to take care of oneself' is translated in French as *se prendre en charge*, which has at once the simple meaning of looking after oneself and, more subtly, connotations of carrying one's own burden, since *une charge* can be used to refer to a weight, responsibility, or other load that one must carry.

I came across a number of cases in which adults and teenagers explained that someone had been sent elsewhere to work specifically to learn how to '*se prendre en charge*' and to avoid '*doing nothing*'. My discussion with T. showed this particularly clearly. He had been institutionally recognised as a former trafficker because he had accompanied a number of adolescents across the border from Zakpota to the mines in Abeokuta, Nigeria. In each case, he claimed, parents had approached him on his visits back to Zakpota asking him to bring their child with him to Abeokuta, so as to avoid them wasting their time unproductively at home and so as to help them learn how to '*look after themselves*' (Interview with T, Southern Benin, 30 August 2007). As the saying in the North of the country has it, '*la poussière des pieds est mieux que celle des fesses*' – '*it is better to have dust on your feet than on your bottom*' (cited Imorou 2009: 7).

That the dust on one's feet may have come from a journey towards mine work is not seen here as inherently problematic. Where the anti-trafficking establishment sees mine work as the kind of work that is *necessarily* unacceptable and exploitative, for the communities with whom I researched, the *nature* of that work and its relation to an individual's capacity is what matters. Thus, in numerous interviews, villagers and former teenage migrants to the mines complained about the nature of the official approach, as did my research assistant, who had been an NGO employee sub-contracted by the state to 'sensitise' locals against such work and movement. '*The state, white people, NGOs, they all come here and say don't let your kids leave for the mines because what they experience there is slavery*', I was told by one village elder (Interview with Artur, Sehere Village, Zakpota, 12 April 2010). '*NGOs call everything slavery ... to stop kids leaving*', Charley, a village head, continued (Interview, Zelele Village, Zakpota, 19 April 2010).

My research showed that labour migration to the Nigerian mines, and the 'trafficking' that has been associated with it, is consistently being translated by the authorities at ground-level as *kanoumon* – the Fon word for slavery. Such is the local frustration with how far removed this designation remains from people's lived experiences of mine work, that when I asked one group of men how they themselves defined the work that teenage boys do in the mines, I received a genuinely emotional round of applause as '*the first person from outside to have ever come here and posed us this question*' (Man in Group 2 Interview, Southern Benin, 12 April 2010). This reaction was not unique. A core component of all the interviews I conducted with former young migrant mine workers and their communities was focused on how they defined their work, how they understood exploitation, slavery, and trafficking and how they viewed the way others view them. Without fail, amid many expressions of frustration, I found that nobody defined work in the mines as *kanoumon*; rather, it is predominantly seen as an acceptable, if challenging, activity that *can*, at times, constitute *afoutame* (exploitation), especially if workers are asked to perform beyond their reasonable capacity. My interview with Petrov was indicative in this regard:

Neil:	*What was the work like there [in Abeokuta]?*
Petrov:	*Very hard, very physically demanding. Plus it meant being under the sun all day. And, when you're sick, the boss moans because it means he's losing money, so if you don't get well again you just get sent home, because you are expensive.*
Neil:	*Do you define this as exploitation?*
Petrov:	*Yes and no, it depends. It can be, if you're asked to do too much, and seeing as wages are so relatively low, but then we do agree to them, and we are looked after. Sometimes we refuse to work too hard as well.*
Neil:	*Did you make friends there?*
Petrov:	*Yes, loads. I could be working here and then I'll have my pal working over there by that tree, and another by that bush. There are loads of kids from around here and they are always bantering whilst working, just as they all eat together at meal times. Daily life there is ok.*
Neil:	*Are you happier here?*
Petrov:	*Yes, but there was also good, there was a good atmosphere. Plus, if you do five or six years and get on with your boss, he will show you the ropes and you can then become a boss yourself.*

(Interview with Petrov, Zelele Village, Zakpota, 10 May 2010)

Migration

Work in, and migration to, the mines of Nigeria is thus viewed in a broadly positive light by these communities. What my interviews revealed, however, is that this understanding of migration transcends the mines; indeed, the decision to migrate is *generally* viewed as a constructive one, in large part because it is the principle vehicle through which people can access paid work and the opportunity that this underpins for self, family, and wider village. This was expressed to me especially when discussing how people perceive the anti-trafficking, anti-migratory messages they hear from the authorities. Below is an example I have extracted from my notes:

Myself, W., about 15 teenage boys, and four or five teenage girls have gathered round in the square at the heart of Atomè Village.

'Do you often hear the message at school that leaving is bad?' Unanimously, like a chorus, they all said yes.

'Why? What do people say?' It is their teachers. Some said occasionally also their parents.

> *They say that 'you can't be better off than where you are from and that you shouldn't leave.'*
>
> *'What do you think when they say this?' 'We think it's rubbish, because we see people coming back with motorbikes and other things and so we know it's untrue'; many nods and agreement. Another boy added that dropping out of school to migrate is not great. But if you are not at school or doing something else then there is absolutely nothing wrong with it.*
>
> *'It's true that you shouldn't leave if you are a pupil', a further boy echoed, 'but if you are doing nothing and your dad says no you can't go away, then that is even bad on his part; even in the summertime you can go and make some money you know.'*
>
> *The girls, through their elected spokesperson, said that as far as they are concerned, if no one helps parents to send their children to school, then it is absolutely fine for kids to leave and find work. They have to!*
>
> *'What do you think of the idea of leaving for work then?' The girls unanimously said that they also thought leaving was good, because there was no money in the village and because the people who leave come back with money, are able to build things for themselves, and then can quickly get married.*
> (Group 12 Interview, Atomè Village, Zogbodomey, 14 May 2010)

These views depend on the concentration of economic opportunity in locations outside of one's home village. Indeed, perceptions of migrant destination centres (Cotonou, Nigeria, or *yovotome* – 'the home of the white man') underline the notion that 'elsewhere' represents a land of opportunity, a place where life is materially richer than 'here', and where one can and must go if one seeks to advance, and not without good reason. The following extract offers a further insight into this perception:

> *'What do you think of when we say "Cotonou"?'*
>
> *Boys:* *'It's a big town that people go to', 'it makes me think of business', 'it's an economic centre and is where people go to find things', 'it is choc-a-bloc full of NGOs and their projects, which we would like to see come here'.*
> *Girls:* *'traffic lights, big roads, things that are there but not here', 'electricity'.*
>
> *'What about "yovotome"?'*
> *Boys:* *'there are only whites over there, that is where they are from', 'many machines', 'big, beautiful buildings', 'cold weather with strong winds'.*
> *Girls:* *'aeroplanes'.*
>
> *'How do you know that money is elsewhere and that leaving will help you get it?'*
>
> *Boys:* *'we see it on the TV, on the serials, the teachers tell us that there is great wealth in Europe, and when a peasant goes away to sell his goods he quickly returns with lots of money.'*

Girls: *'we have relatives who are well-off in Cotonou and when they come back here to visit we can see that they are living much better than we are.'*

(Group 12 Interview, Atomè Village, Zogbodomey, 14 May 2010)

These perceptions are evidently quite acute. Despite the establishment claim that 'home' is where the young should be, it is clear that economic wealth is concentrated in large urban centres rather than the countryside, in Nigeria rather than Benin, and in the West rather than sub-Saharan Africa. Such notions of 'elsewhere' as a land of opportunity are not confined only to my interviewees, or indeed only to the rural poor. In fact, large-scale work by the UNDP in Benin on rural and urban perceptions of poverty and socio-economic well-being echoes what I found in my interviews. In one study conducted with a series of rural and urban communities in the South of the country, both populations saw themselves as relatively poor, both saw one's own location as synonymous with negative economic prospects, and both saw various alternative destinations as embodying opportunity (PNUD and MPD 1995, 1996).

Given such understanding, various of my interviewees repeatedly articulated that they believed the migration of both young and old served as a tool for local development. In one particularly telling example, two female elders railed passionately against the anti-trafficking policy establishment:

'What do you think of the message that young people shouldn't leave the village?'

Those who tell us this are those who hold back the development of this village! It is a terrible message! And they give us nothing in return. The NGOs come here but they bring nothing with them! (Woman 1)

'Why do NGOs and the government do this and say these things?'

They don't want people to leave the village because they don't want to see us go and develop elsewhere instead of here. Fair enough, but their words are useless, because they bring nothing. (Woman 2) (Group 6 Interview, Atomè Village, Zogbodomey, 28 April 2010)

Though expressed with less vitriol, the words of another adult respondent offer a similar analysis:

There is just no money in Benin in general. Nigeria has lots of opportunities and so people prefer to go there, work or buy things and then come back. This helps the village. For example, if one child leaves and comes back with a generator, that means the whole village can have access to light at festival times. (Man in Group 11 Interview, Tenga Village, Zogbodomey, 14 May 2010)

The centrality of cash

The above examples demonstrate the importance of cash and access to it in the decisions of young people, or those who decide on their behalf, to migrate for work. That cash should be so important was further underlined when I examined with the communities what was meant by the Fon word *ya* – 'poverty'. Though poverty frequently featured as an answer to the question '*why do young people leave here?*', further discussion revealed that, in contrast to the anti-trafficking notion of poverty as destitution, 'poverty' here simply means a lack of the cash necessary to '*evolve*'. When I asked if 'poverty' meant starvation and whether 'poor' people remaining in the village would go without food, most people responded with an amused and resounding '*no*'. Pointedly, one man explained that '*people don't die of hunger here*', while another declared that '*even when there are no fish, people don't starve ... people wouldn't let you starve, it's just that there's no money*'.

Such linguistic digging was not necessary with all my interviewees. As the above examples testified, the importance of accessing money or material opportunity is implicit in most migrant narratives. In many others, it is voiced explicitly as an inevitable part of life in a market economy:

> *There is nothing in the village, there is no work. Parents are obliged to let their kids go and when kids decide themselves to leave, parents are obliged to accept. When they go, kids at least make some money; they at least send some back to us. We understand the "don't leave" message, but we can't eat their words can we?* (Man during Group 3 Interview, Tenga Village, Zogbodomey, 16 April 2010)

> *I was in Nigeria working in the mines until only five months ago. If I'm not in Nigeria, I'm in Savè, because at least that gives me some money. I remember that one NGO came and said don't leave, promising to bring money for those who stayed, but they never did. Those who stayed had been tricked and were really sad, especially in comparison to those who did leave, and who made some money.* (Teenager during Group 3 Interview, Tenga Village, Zogbodomey, 16 April 2010)

The need for alternatives

Village-level understanding of what policymakers *should* be doing to protect the young from exploitation (or 'trafficking'), therefore, differ markedly from those that prevail within the policy-making system. When I asked young migrants and their communities the question, '*what would you like to see the authorities do?*', two responses consistently made themselves heard:

1. Provide us with economic alternatives to labour migration
2. Ensure that those who do migrate can work in safe conditions

Key for these communities is the ability to access the money that underpins their individual and collective life projects. If possible, they would like to have the option to do this at home, but where they cannot, they at least wish to be able to work in safety. In the absence of policies that respond to these needs, communities both ignore and subvert the authorities and their attempt to control their work and migration. The following extracts from my interviews offer an indication of this fact:

> *'Do you pretend to the NGOs and government, saying one thing to them and doing another?' There was lots of laughter amongst those that understood my question. Everybody said yes, they do.* (Interview with Group 5, Zelele Village, 26 April 2010)

> *'So you just pretend to the authorities then?'*

> *Yes, of course. We say 'sure, we won't leave' in the hope that they'll bring us something.*

(Interview with Artur, Sehere Village, Zakpota, 12 April 2010)

'How do you go about getting around the authorities?'

They said the state has set up village committees all over the place, but these are corrupt. We can easily turn them and take kids away no problem. There are also many paths that you can take towards and across the border and the state has no idea about them all. The police sit there and guard the ones they know about and so we just take the others. (Interview with Group 2, 12 April 2010)

Unless the authorities are prepared to alter their draconian and abolitionist approach to work, migration, and trafficking, little is likely to change in this regard.

Cotton and the political-economic threads of poverty

In light of the importance of money within the decision for the young to engage in labour migration, an obvious question is: what is the cause of this lack of money? Though there is not space to argue the case definitively here, my research with young migrants and their communities and related political-economic analysis suggests that it comes from the declining household income from cotton, the exposure Southern Beninese households have to commodity price volatility, and the US cotton subsidies that are a causal factor of the recent international depression in cotton prices.

At the national level, cotton accounts for around five per cent of Benin's GDP and almost 40 per cent of the country's export receipts (OECD 2005: 20). It is the country's major cash crop, and in villages including those in which I researched, there is no comparable way for many farming families to access money without migrating, since the climate is perfectly adapted to cotton, substitutes do not benefit from similarly well-organised agricultural markets, and there is minimal industrial development.

When discussing its relation to household livelihoods, farmers and state agricultural agents argued that when cotton works well, life is good, the country has money to spend, families have money to build, children go to school, and investments are made. One elderly woman explained, '*Before prices stayed low, all the young people earned some cash. They worked the fields and the old owned the land and paid them*' (Group 6 Interview, Atomé Village, Zogbodomey, 28 April 2010). A local agricultural official echoed this assessment, saying,

I can tell you, [before prices depressed], farmers used to look after us, they were so well off. They had money, people built houses, wells, had weddings and ceremonies, kids had money and went to school, people bought tractors. (Interview with Donald, Regional State Agricultural Official, 26 April 2010)

In contrast to this, since the onset of price depression in the mid-1990s, small-scale household projects have been put on hold, more and more family members have had to enter the wage market, and ultimately many more have migrated, to Nigeria or to the cities. When asked, '*Is it true then that young people move more now that cotton returns are low?*', one of the most indicative responses I received was:

They wouldn't move at all if they were still high. They'd all be in school. (Interview with Cliff, Sehere Village, Zakpota, 7 April 2010)

Research suggests that these lower prices are a result of more than 'natural' international price fluctuations. Arguably, they derive in part from US cotton subsidies (Alston *et al.* 2006). In a World Trade Organisation (WTO) case brought by Brazil, the WTO recognised that subsidies depress world (and thus household) prices (Sumner 2007: 7), since they de-couple US cotton production from world cotton prices, such that US cotton growers fail to respond in economically rational terms to lower prices by reducing production, and instead maintain or even increase output levels, which further depresses the market (Cross 2006; Eagleton-Pierce 2010; Fadiga *et al.* 2004; Fok 2002; Oxfam 2002; Sumner 2007).

When I tried to address this issue with anti-trafficking policymakers in Benin, the responses I received ranged from blank incredulity to the admission that engaging on such ground was politically impossible. For the most part, institutional interviewees failed to make the link between the political economy of subsidies and reduced household income. Only one did so without my prompting. With the others, 'poverty' was presented as a causeless reality. When I suggested that anti-trafficking policy should tackle the link between subsidies and poverty, my respondents were clear that this would not happen. JayJay, for example, explained to me:

> *We can't go beyond the limits of a country with our work. This is a national structure, it's a national delegation. We structurally cannot go beyond borders.* (Interview with JayJay, Senior Donor Government Representative, Based in Benin, 2 April 2010)

Rose, a senior international organisation employee working at country level in Benin, curtly explained that her institution simply couldn't look at structural issues because it was '*politically unacceptable*' (Interview 29 September 2009), while Matt, a crucial figure in one donor country's head office, said that he intended to try and raise the issue, but was doubtful of success because there were '*big interests to fight*' (Interview, 16 September 2009). Perhaps most tellingly, a senior government representative based in Benin, had this to say:

> *The foreigners here in Benin know that their policies cause poverty and trafficking. Many of them even want to change it, but they can't. Politicians aren't going to change these policies just because a few kids are trafficked. In their reports they just say "poverty is the cause". They say the poor national use and distribution of resources is a problem. It is organisational hypocrisy, we know it's bad, everyone knows, but what can we do? It's very difficult to change the macro-structure.* (Interview, 17 February 2010)

Conclusion

Working on the macro structure is, however, necessary if anti-trafficking policy in Benin is likely to be anything other than an obstacle that poor communities must navigate as they pursue their life goals. Over the course of this paper, I have demonstrated that the understandings of child work, migration, and exploitation that predominates within the communities targeted by anti-trafficking policy neither reflects the empirics of young migrant lives nor responds to the way young migrants see their lives. Unsurprisingly, therefore, a major consequence is that the communities targeted by anti-trafficking strategies actively resist and undermine those strategies.

If policymakers wish to see a change in this situation, greater participation and greater politics are needed. At the macro-level, it is essential that they address the very real, very tangible political-economy underpinnings of the lack of money of that constitutes the main backdrop to the migration and work of the young in this part of the world. At the micro-level, they need to engage with communities in developing economically viable alternatives to labour migration and pathways to ensure that those who do migrate for labour are able to do so in safety. Measures could include, for example, the expansion of labour inspectorates and the involvement of employers in establishing labour standards. They could also involve institutions speaking with one voice about the negative effects of US cotton subsidies, while fostering direct investment in supplementary industries in zones from which migrants come.

If they do not do this, little is likely to change at the level of the lived realities of young labour migrants and their communities. Indeed, it might be suggested that policy's major accomplishments will be legally and institutionally to entrench dominant norms relating to acceptable childhoods and to naturalise political-economic injustices. Regarding the latter, Ferguson (1990) warned that policy efforts that abstract lived realities from their political-economic contexts lead ultimately to the normalising and perpetuation of those contexts. In the case of youth labour migration from the Beninese cotton belt, by focusing on the symptom, namely movement, instead of the cause – the subsidies – anti-trafficking policy detracts critical attention from the conditions leading to what it seeks to prevent. Regarding this, we should recall Boyden's (1997) warning: policies that seek to intervene in the *way* families live, rather than on the conditions which structure *why* they live this way, contribute to the propagation of notions of 'normality' and 'deviance'. This in turn leads to the criminalisation of those who do not adhere to the accepted norm – as has been the case with the migrant-sending families I worked with in Benin.

Notes

1. All names of individuals, institutions, and villages have been either changed or omitted to protect the identity of informants. Unless otherwise indicated, all translations from French are my own.
2. Though I consider 'children' and 'childhood' to be situational concepts varying according to time, place, culture, and social structure, amongst other things, the institutional literature and policy I am examining takes 'children' to be those under the biological age of 18 and 'childhood' thus to be the period before one reaches that age. Since my research reflects on the appropriateness of these institutional understandings and policies, I have decided to engage with them on their own terms, and have therefore used the same criteria throughout this study. In my study, however, I focused almost entirely on adolescent migrants, and the word 'child' will be used interchangeably with 'youth' unless otherwise specified.
3. The Etireno was a Nigerian trawler used by a gang of people-smugglers to transport illegally Beninese and other West African children to work in Gabon. After a complex series of events, the ship and its passengers were found stranded at sea in difficult conditions for a number of days. This prompted the world's media to descend on the Gulf of Guinea and lead to a flurry of high-profile reports on 'the slave ship' that heralded the uncovering of Benin's 'modern-day slave trade' (see, for example, *The Independent*, 18 April 2001).
4. This treaty is officially titled the 'UN Protocol to Prevent, Suppress and Punish Trafficking in Persons, Especially Women and Children, Supplementing the Convention on Transnational Organised Crime' (UN 2000). It defines 'trafficking in persons' as '*the recruitment, transportation, transfer, harbouring or receipt of persons, by means of the threat or use of force or other forms of coercion, of abduction, of fraud, of deception, of the abuse of power or of a position of vulnerability or of the giving or receiving of payments or benefits to achieve the consent of a person having control over another person, for the purpose of exploitation*'.

References

Alber, E (2011) 'Child trafficking in West Africa', in A. M. Gonzalez *et al.* (eds.) *Frontiers of Globalization: Kinship and Family Structure in Africa*, London: Africa World Press.

Alston, J. *et al.* (2006) 'Impacts of Reductions in US Cotton Subsidies on West African Cotton Producers', Washington, DC: Oxfam America.

Boyden, J. (1997) 'Childhood and the policy makers: a comparative perspective on the globalisation of childhood', in A. James and A. Prout (eds.) *Constructing and Reconstructing Childhood: Contemporary Issues in the Sociological Study of Childhood*, London, UK: Falmer Press.

Cross, K. H. (2006) 'King cotton, developing countries and the "Peace Clause": The WTO's US cotton subsidies decision', *Journal of International Economic Law* 9 (1): 149–95.

Eagleton-Pierce, M. (2010) 'Uncovering Symbolic Power: Power Analysis, Southern Countries, and the World Trade Organization', unpublished DPhil thesis, University of Oxford, UK.

Fadiga, M, S. Mohanty, and S. Pan (2004) 'The Impacts of US Cotton Programs on the West and Central African Countries Cotton Export Earnings', paper presented at the American Agricultural Economics Association Annual Meeting, Denver, Colorado, USA, 1–4 August 2004.

Ferguson, J. (1990) *The Anti-Politics Machine: 'Development', Depoliticization and Bureaucratic Power in Lesotho*, Cambridge, UK: Cambridge University Press.

Fok, M. A. C. (2002) 'Préserver un Futur au Coton Africain: la Revendication Légitime d'Équité dans les Échanges Mondiaux Ne Sera Pas Suffisante', Geneva, Switzerland: UNCTAD.

Hashim, I. and D. Thorsen (2011) *Child Migration in Africa*, London, UK: Zed Books.

Howard, N (2011) 'Is child placement trafficking? Questioning the validity of an accepted discourse', *Anthropology Today* 27 (7): 3–7.

ILO (International Labour Office) (1973) *Convention Concerning Minimum Age for Admission to Employment*, Geneva, Switzerland: ILO.

Imorou, A.-B. (2009) 'Children's And Young People's Mobility: A Study Of Tactics And Strategies And Involvement In Managing Their Social Trajectories', Working Paper presented at a workshop on Child and Youth Migration in West Africa: Research Progress and Implications for Policy, Ghana, Accra, 9–10 June 2009.

OECD Sahel and West Africa Club Secretariat (2005) 'Economic And Social Importance Of Cotton Production And Trade In West Africa: Role Of Cotton In Regional Development, Trade And Livelihoods', Paris, France: OECD.

Oxfam (2002) 'Cultivating Poverty: The Impact of US Cotton Subsidies on Africa', Oxfam Briefing Paper', Oxford, UK: Oxfam.

PNUD (Programme des Nations Unies pour le Développement) and MDE (Ministère du Développement Rural) (1995) *Études sur les Conditions de Vie des Ménages Ruraux (ECVR) : Perception des Dimensions de la Pauvreté, du Bien-Être et de la Richesse dans les Quartiers Pauvres en Milieu Rural au Bénin*, Cotonou, Benin: PNUD.

PNUD and MDE (1996) *Études sur les Conditions de Vie des Ménages Ruraux (ECVR) : Perception des Dimensions de la Pauvreté, du Bien-Être et de la Richesse dans les Quartiers Pauvres en Milieu Urbain au Bénin*, Cotonou, Benin: PNUD.

Sumner, D (2007) 'US Farm Programs and African Cotton', IPC Issue Brief 22', Washington D.C. USA: International Food & Agricultural Trade Policy Council.

The spatialisation of child protection: notes from the occupied Palestinian territory

Jason Hart

This paper considers the employment of spatialised practice within child protection efforts as pursued by humanitarian agencies. Starting from a brief overview of the genealogy of enclosure and separation within both humanitarianism and in relation to childhood, examination is then made of spatialisation in the setting of the occupied Palestinian territory. It is argued that in a situation of occupation, the spatialised approach entails numerous problems that are both practical and political in nature.

La spatialisation de la protection de l'enfance : notes du territoire palestinien occupé
Cet article traite de l'utilisation de pratiques spatialisées dans le cadre des efforts de protection de l'enfance par les agences humanitaires. En commençant par un bref aperçu de la généalogie de l'enfermement et de la séparation à la fois dans le cadre de l'humanitarisme et par rapport à l'enfance, un examen est ensuite effectué de la spatialisation dans le contexte du territoire palestinien occupé. Il est soutenu que, dans une situation d'occupation, l'approche spatialisée donne lieu à de nombreux problèmes qui sont de nature pratique et politique.

A espacialização da proteção infantil: notas do território palestino ocupado
Este artigo considera o emprego de práticas de espacialização nos esforços de proteção infantil implementados por agências humanitárias. Iniciando com uma breve visão geral da genealogia do isolamento e separação dentro do humanitarismo em relação à infância, é feito então um exame da espacialização na criação do território Palestino ocupado. Argumenta-se que em uma situação de ocupação, a abordagem de espacialização envolve vários problemas de natureza prática e política.

La espacialización de la protección de la niñez: notas desde el territorio palestino ocupado
Este ensayo analiza cómo las agencias humanitarias utilizan el método de espacialización en sus acciones destinadas a la protección de la niñez. El análisis comienza con un breve repaso a la genealogía del cercamiento y del aislamiento en los ámbitos humanitario y de la niñez, seguido por una revisión de la espacialización en el contexto del territorio palestino ocupado. El ensayo sostiene que, en una situación de ocupación, el método de espacialización acusa varios problemas de índole práctica y política.

Introduction

The contemporary discourse and praxis of child protection in settings of humanitarian emergency reveal a strong reliance upon spatialised thinking. Recent years have witnessed a plethora of space-based strategies intended to ensure the relative safety of young people at risk of harm as a result of armed conflict and natural disaster. 'Child-friendly/safe spaces', 'children as zones of peace', 'protective environment' are just some of the terms coined by protection practitioners in recent years. Each implies a direct, causal relationship between the circumscription of space and the better protection of the young.

The spatialised approach to child protection is predicated upon assumptions about the separability of children from the wider environment, not only physically, but in cognitive and emotional terms as well. This in itself reveals the dominance of a mode of thought that has been evident within Europe and North America since at least the nineteenth century, growing stronger ever since. The important question here is whether a spatialised approach to child protection, which is informed by such thinking, is effective across the diverse settings where humanitarian organisations operate. This essay is a modest contribution towards answering such a question, drawing upon research conducted in a single location: the occupied Palestinian territory (oPt). My aim is to illustrate the challenges to the spatialised approach in this setting, paying attention to the specific additional motivations for international and UN agencies working there. While the findings from the oPt may be highly specific, I hope that the mode of enquiry followed in this paper may prove helpful for examination of child protection practice in other locations.

I begin by offering a tentative genealogy of thinking about enclosure and separability as it has evolved within humanitarianism generally and child protection more specifically. My aim is to explicate the historical and cultural contingency of the spatialised thinking that informs much current child protection practice.

Spatialisation within humanitarian practice

The rationale for bringing people together within a defined area for their better protection is not difficult to grasp. Concentration of a vulnerable population enables the delivery of food, medical care and basic services in situations where the essential materials for survival are otherwise hard to obtain. Moreover, agencies concerned for the protection of people who may be especially vulnerable are more able to observe and, if necessary, intervene on behalf of an enclosed rather than dispersed population.

Enclosure of a population in a clearly marked space can, however, create its own risks. Such containment may become a virtual prison or, at the least, a hindrance to livelihood strategies and the conduct of conventional social relations (Tefferi 2008). Disease can spread more easily in a congested space with limited infrastructure, and various risks associated with the enforced mingling of disparate groups can arise. In many humanitarian settings, encamped populations have been rendered a highly visible target for attack or exploitation (e.g. Lischer 2005). Flight, dispersal, and hiding may, in some situations, be a more effective means of self-protection. Over half the 10.4 million refugees 'of concern' to the UNHCR in 2011 were believed by that organisation to live in urban areas: a fact that partially reflects the preference to avoid encampment.

Given that the benefits of enclosing a population within a defined space are not automatic, and that dispersal has often proven advantageous, we should look beyond intrinsic factors to understand the growing employment of a spatialised approach to protection since the mid-twentieth century. Here the recent history of refugee encampment – the most obvious mode of spatialised humanitarian practice – is instructive. The evolving use of the displacement camp in the period since World War 2 has been driven by more than concern for refugees' better protection:

> *The segregation of nationalities; the orderly organisation of repatriation or third-country resettlement; medical and hygienic programs and quarantining; "perpetual screening" and the accumulation of documentation on the inhabitants of the camps; the control of movement and black-marketing; law enforcement and public discipline; and schooling and rehabilitation were some of the operations that the spatial concentration and ordering of people enabled or facilitated.* (Malkki 1995: 498)

As this range of functions indicates, encampment serves the needs of authorities as well as those of the displaced population themselves. Encampment also enables efforts to impart norms of behaviour and disposition that are taken for granted as beneficial by (mostly Western-originating) organisations.

In the various functions which the camp enables, we see the relationship of organisations involved in establishing and managing such camps, such as UNHCR, to the agendas of host governments and donors. Common to the agendas of host and donor governments is concern to manage the movement of peoples who might otherwise make unwelcome claims on national resources or pose problems for social cohesion. Encampment also offers agencies charged with the welfare of displaced populations the opportunity to demonstrate their own efficacy, thereby strengthening their efforts to secure further funding since, as Harrell-Bond has noted, '*[t]o attract money, refugees must be visible*' (1986: 8).

Moreover, by containing a displaced population within a bounded space away from urban centres the potential of camp residents to attract media attention to their cause might be constrained. This may be important not only for host governments seeking to minimise negative international publicity but also for humanitarian agencies keen to avoid involvement in political debate that endangers their own claims of neutrality. In short, encampment offers a number of advantages to humanitarian organisations, the governments that host them, and the donors that provide their funding which extend far beyond the better protection and support of resident populations. The continued use of camps and other enclosed facilities cannot, therefore, be seen as an inevitable response to a protection crisis.

Spatialisation of childhood

The spatialised approach to child protection should be understood in part as the extension of practices commonplace in the humanitarian field more generally. As such, it is motivated by similar considerations as evident in encampment discussed above: considerations that include the institutional needs of implementing organisations, the political agendas of host nations and donors, and the imposition of certain (usually Western-originating) norms upon disparate (usually non-Western) populations.

In addition, there are ideas and practices relating to children within Euro-American societies that prompt the use of enclosed space for the protection of young people living in settings of political violence and natural disaster. The separation and enclosure of childhood itself from the larger 'adult world' has been a defining characteristic of many such societies since at least the late nineteenth century, becoming ever more evident in the decades since. Thus, current spatialised child protection efforts need also to be understood in relation to this recent history within the settings from which contemporary humanitarianism emerged.

The proposition that childhood should be conducted away from the contaminating effects of the 'adult world' is particular to a tradition of European thought that finds its genesis in the Romanticism of Jean Jacques Rousseau (1969 [1762]). However, it was not until the late nineteenth and early twentieth centuries that such separation became a social reality amongst the masses. On one hand, children were withdrawn from industrial work, partly due to moral concern but also as a result of various material factors (Zelizer 1985). On the other hand, the establishment of mass, compulsory schooling was driven by a mixture of motives that included the wish to curtail delinquency amongst unemployed young people from the 'dangerous classes', the surveillance needs of the state, and an aspiration to give children a 'proper childhood'. A key feature of such a childhood was its separation from 'adult life':

> *Childhood and adulthood ... became almost opposites of one another. If adults were burdened with responsibilities, children should be carefree. If adults worked, children should not work. If adults had to live in towns, children were entitled to contact with nature.* (Cunningham 1995: 160)

During the twentieth century, particularly in the latter decades, the separation of children from adults and their enclosure within 'appropriate' spaces increased considerably. Sociologists have drawn attention to the 'islanding of children' within many Euro-American societies, whereby '*children have been systematically excluded from the former mainlands of urban and suburban existence, especially the streets and other public spaces*' (Gillis 2008: 316). The reasons prof-

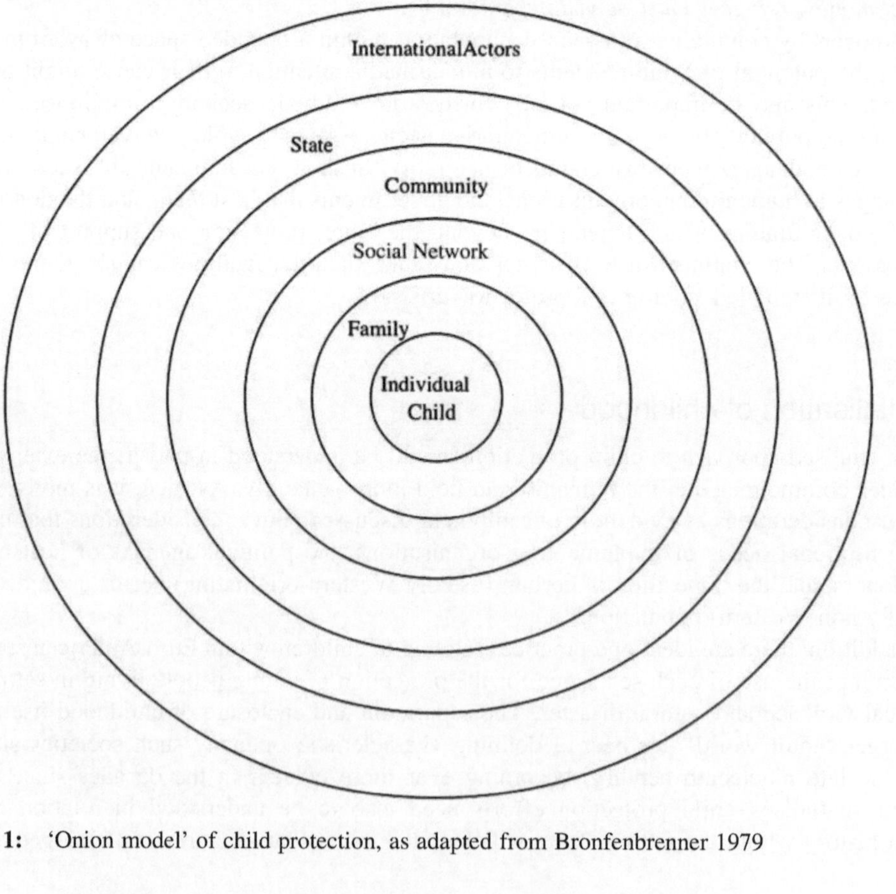

Figure 1: 'Onion model' of child protection, as adapted from Bronfenbrenner 1979

fered for the increasing containment of children within designated spaces are complex, relating in the view of some to the conditions of postmodernity and a sense of rootlessness amongst adults (Jenks 1994). The spate of moral panics witnessed in recent years and the emergence of discourse around 'stranger danger' are also implicated in such enclosure. Ignoring both the lack of evidence that children are at any greater risk than in previous times as well as the proof that they are most likely to experience harm within the home, such 'islanding' continues.

In the late 1970s, social psychologist Bronfenbrenner suggested a model of child development that captured the spatialised thinking around childhood extant at that time and increasingly since. This 'ecological model' depicts the child at the centre of a series of concentric circles each of which represents a level of social interaction at an increasing distance: from 'family' at the most immediate level, to 'state' and other macro level duty bearers at the outer circles.

Since the early 1980s this model has been used for diverse purposes that include child participation and child protection. In respect of the latter, a key task for each of the identified duty bearers is to secure the space implied by the gaps between the lines through removing the threats to children.

The ubiquity of Bronfenbrenner's model not only reflects existing assumptions but has also served to naturalise further the spatialised thinking around child protection. The influence of this model can be seen in the notion of 'protective environment' employed, often implicitly, within social work practice in Western Europe and North America (e.g. Jack and Gill 2010). The significant role of social work expertise in the development of child protection in emergency settings has been an important means by which the model of Bronfenbrenner has become central to ways of thinking in this field. Permutations of the basic figure appear in numerous handbooks and presentations seeking to guide current practice (e.g. Ager 2011). According to the model, family and community constitute the primary duty bearers, securing an immediate locale within which protection is best achieved. A major focus of humanitarian organisations is to strengthen the protective functioning of these duty bearers through interventions that typically include psychosocial work, family reunification, and creating referral networks to alert duty bearers at outer levels when problems occur within the family or community. As we shall see, in the oPt the basic topography of this model is brought into question. However, before turning to consideration of the protection of Palestinian children, we first need to consider the spatialisation of child protection practice in humanitarian efforts generally.

Spatialisation of child protection in emergencies

In addition to promoting the protective function of household and immediate neighbourhood child-focused humanitarian organisations are engaged in an evolving range of initiatives that entail identification or creation of additional safe space. An obvious example of this is the 'child friendly space' – a notion invoked by many humanitarian organisations. According to the understanding of World Vision, for example, a 'child-friendly space' is:

> *. . . a structured and safe place where children and youth meet other children to play, learn competencies to deal with the risks they face, be involved in some educational activities and relax in a safe place. It gives the children a sense of safety, structure and continuity that provides support amidst overwhelming experiences.* (World Vision International 2005: 1)

In some locations agencies are active in promoting the notion that the space of protection should be co-terminus with the physical being of individual children. This can be witnessed in advocacy campaigns to treat 'children as zones of peace' (CZOP). In addition, the territory to be kept apart from the surrounding conflict may be defined in terms of the institutional space that children are assumed to inhabit routinely – most obviously the school. Thus, for example, in Nepal, a campaign focussed around CZOP has led to a further campaign of SZOP – 'schools as zones of peace'. The stated aim of the SZOP campaign is '*to keep all schools free from political and other interference; discrimination, violence and corporal punishment; neglect and exploitation*' (CZOPP 2011). This speaks to a range of aims, only some of which relate directly to children's needs for protection from the threats resulting from armed conflict. In the Nepali context, for example, discrimination in schools on the basis of caste is a long-standing issue that far pre-dates the Maoist-led 'People's War'. The inclusion of an aim to end discrimination (as well as corporal punishment, neglect and so on), illustrates the ways that the securing of space in the name of children's protection often goes hand-in-hand with the pursuit of other aims that have little to do with risks arising from conflict. As for the general population in refugee camps so for children: enclosure can serve a range of functions beyond immediate protection.

Spatialisation of child protection in the oPt

The setting

Discussion about child protection in the oPt is informed by fieldwork conducted in the West Bank and East Jerusalem in 2009, supported by prior qualitative research conducted on different occasions over several years, by subsequent interviews and by extensive engagement with practitioner and academic literature. The research in 2009 that Claudia Lo Forte and I undertook focussed specifically on the role of organisations in protecting Palestinian children (Hart and Lo Forte 2010).

There is not the space here to offer a detailed description of the oPt as a setting of political violence, for which a plethora of excellent publications exist. The intention, rather, is to offer sufficient contextual information to make clear the implications of employing a spatialised approach to child protection.

Within the media and in popular political debate in the West the notion of an 'Israeli-Palestinian conflict' is commonplace. UN and international humanitarian organisations similarly invoke the conflict paradigm connoting enmity between two equivalent forces fighting across clear lines of battle. These agencies respond with a range of measures drawn from a basic toolkit intended for all settings of humanitarian emergency – both natural disasters and armed conflict. Such conceptualisation obscures the fundamental point that the risks faced by Palestinian children arise not because of conflict as conventionally understood but as a result of Israel's occupation. In the oPt, the notion of battle-lines from which children may be kept distant has little purchase given the reach of the Israeli authorities and the activities of extremist settlers living in the midst of the Palestinians.

Occupation in this setting is not static: the matrix of control that Israel exerts over East Jerusalem and the West Bank enables the progressive appropriation of land and resources. As more is taken for the construction of settlements and in the name of security or environmental protection, so the Palestinian population find themselves confined within shrinking pockets of land that can be sealed through checkpoints, curfews, and the closure of gates in the 'separation wall'. This process of colonisation and accompanying enclavisation is evidenced in numerous statistics. For example, in East Jerusalem some 23,378 dunams (ca. 5,750 acres) of Palestinian land were appropriated for Israeli (Jewish) use in the period 1967–91, while, in the period 1967–2010 more than 13,000 Palestinian Jerusalemites were forced to leave the municipality through revocation of their residency rights. Almost 500,000 Jewish settlers have moved into settlements in East Jerusalem and the West Bank. Additionally, Palestinians are routinely denied permits to build in areas under Israeli civil control (all of East Jerusalem and 59 per cent of the West Bank), leading to the demolition of homes necessarily built without permits (over 300 in the period 2007–11). As of October 2010 there were 99 fixed checkpoints, 62 of which were internal to the West Bank hindering the population from moving between areas of Palestinian civilian life. So-called 'flying checkpoints' further compound the obstacles to movement, which are especially severe for males aged 16–35 (for all statistics see www.btselem.org). The evident intention of such practices is to concentrate the Palestinians within enclaves around the main cities and towns, while securing land for construction and agricultural projects from which Israeli Jews derive principal benefit (Dajani 2005). In short, Israel has been pursuing policies of enclosure in a dedicated manner for decades.

Into this setting of occupation and colonisation, humanitarian organisations have brought ways of thinking and a set of practices around child protection that is focused to a considerable extent upon separation and enclosure. There are also activities that do not entail a spatialised dimension: some organisations seek to address the causes of threat to children's lives in a holistic manner that relates to the wider political-economic environment. Overall, however, such efforts have been less routine and less concerted than those that employ a spatialised approach. The forms that spatialisation takes in the oPt, and the implications it entails in light of Israeli policy will both be discussed. In addition I shall also consider the institutional agendas that are served by a spatialised approach to child protection in the oPt and which may enable, in part, their perpetuation in spite of the problematic implications.

Spatialisation

In the oPt spatialisation takes many forms, some more explicit than others. These might be categorised as: (1) efforts to strengthen the presumed protective spaces of family and community; (2) efforts to transform existing institutional spaces, making them safer and more child-friendly; and (3) efforts to create social spaces that serve a protective function. I shall briefly describe each of these in turn, offering observations about the particular implications of each in the context of occupation.

(1) *Efforts to strengthen the presumed protective spaces of family and community*. In recent years a large proportion of child-related programming in the oPt has focused on psychosocial interventions of various kinds. For some organisations, psychosocial programming has become synonymous with child protection programming, while for others the two remain partially distinct. Under the psychosocial label, several organisations focus on efforts intended to enable families to support children better by aiding caregivers to manage their own stress or by imparting techniques intended to equip caregivers to support children through traumatic times more effectively. In the event that families and communities fail to care appropriately for individual children, referral networks have been created. Part of their function is to identify 'at-risk' children and ensure that proper institutional support is given.

The popularity of such measures amongst international and UN agencies, and the governmental donors that fund them can be attributed to at least two factors. First, these are standard responses employed not only in other settings of emergency but by social workers in many Western countries as well. Undoubtedly efforts to support families through psychosocial programming are important and valued in the oPt, and a growing number of local organisations are involved in their delivery. However, the immense investment of UN and international organisations in such programming, particularly in light of local competence, should alert us to other factors behind this focus. Certainly, psychosocial programming enables agencies to achieve (and demonstrate) some degree of efficacy by limiting the goal to coping and reducing harm *within* the conditions of occupation, rather than challenging the source of the problems. While not all domestic violence can be attributed to occupation there is no doubt that caregivers are commonly placed under unbearable strain and may express their frustration in behaviour that is abusive or neglectful. As one 12-year-old girl explained:

> *Now if you experience the occupation, you have someone in prison or you don't know where they are ... you will feel angry, stressed, you feel you cannot protect yourself; that will affect how you respond to the environment around you. So if my dad who's a taxi driver, if the soldiers take his ID or his driving license, when he comes back home he will be angry and doesn't want to listen to us.*

Agency staff interviewed in 2009 readily acknowledged the negative impact of occupation upon the domestic realm. Nevertheless, in practice the focus of interventions is on the enhancement of coping strategies or upon identifying children 'at risk'. This I would argue follows, in part, from a conceptual separation made by many practitioners in the oPt between 'external' or 'occupation-related' issues, on one hand, and 'internal' or 'traditional' issues on the other. The employment of such a dichotomy contributes to a sense that these can be addressed in isolation from one another and that behaviours within the space of the home can be improved without challenging the conditions of occupation.

For some Palestinian children – particularly those living in refugee camps, in neighbourhoods close to extremist settlers, and in the Jordan Valley – the boundedness of the domestic/community realm is fragile. For these children the separability of the space of family and community from the larger political setting is questionable given the experience of Israeli soldiers entering their homes at will, wrecking property and arresting them or family members. Moreover, homes themselves are routinely destroyed on the grounds that they were built without (unobtainable) permits. Between 2007 and 2011 nearly 1,000 children were made homeless in this way (www.btselem.org). Assumptions about the ability of family and community to secure the most immediate spaces of children's lives and afford them protection, which lie at the core of the spatialised approach, are clearly questionable in light of such experiences.

(2) *Efforts to transform existing institutional spaces, making them safer.* At the global level considerable attention has been given to the school as a potentially protective space for children living in situations of emergency. Initiatives such as 'healing classrooms' involving the International Rescue Committee or 'Schools as Zones of Peace' mentioned above seek to realise this potential through advocacy, community involvement and capacity building. In the oPt, the 'Protective Sphere' project has pursued a similar objective of transforming the existing school space into an environment more conducive to children's well-being. This is a project of Save the Children UK (SCUK) partnered by a local NGO, with funds from the European Commission. As SCUK describes the project:

> *... we're building a 'protective sphere' to reduce violence in schools and in the wider community. We're carrying out activities in 15 schools ... benefiting 7,500 children. Working with teachers, children and their parents, we provide training on children's rights and alternatives to corporal punishment. Other activities include setting up school-based education committees (SBECs) in each of the 15 schools we support, training teachers and parents on how they can make children safer.* (2009: 3)

Working to create a school environment in which children are less fearful of violence from teachers or their peers is an important endeavour in any setting. However, in the oPt, the achievement of this aim rests on the assumption that violence within the school can be countered separately from the violence of occupation. As a child quoted in a report by SCUK/SC Sweden observed:

> *Teachers themselves are deprived of their rights and they are stressed. That's why they practice violence.* (SCUK and SC Sweden 2004: 22)

As with the domestic/community realm, school premises and other institutional settings frequented by children, including clinics, are commonly subject to threatening actions by the Israeli authorities. Leaving aside the attacks on school buildings and health facilities during Operation Cast Lead in Gaza during winter 2008–09, Palestinian children have been routinely put at risk during military operations and incursions or as a result of attacks by settlers. Some schools have been targeted repeatedly, such as al-Qurtuba girls' school in Hebron which lies amongst neighbourhoods inhabited by some of the most extremist settlers in the oPt. In East Jerusalem many children are obliged to learn in overcrowded, unsuited and, in some cases, unsafe buildings due to a shortage of facilities and Israeli constraints on creation of the needed infrastructure.

For some Palestinian children the greatest danger they face is not within the school premises but on the journey to and from school. In a number of locations this is especially acute. For example, in the South Hebron Hills children from the villages near the settlement of Ma'on and the outpost of Havat Ma'on have been physically assaulted by settlers as they walk to their school in al-Tuwani (CPT & OD 2010). More broadly children risk verbal and physical abuse, and may witness the humiliation of parents and other family members on their journey through Israeli checkpoints. One consequence is drop-out, as a 14-year-old girl explained:

> *Our parents send us out early in the morning and don't know if we will return. There are girls who gave up school because of the trouble on the way and the cost of the journey.*

For Palestinian children the challenges of travelling to school and other institutions offering needed services, suggest that the focus upon safe space alone is inadequate. Also needed in this setting are concerted efforts to achieve safety in movement. In many emergencies, people prefer to stay in a limited but relatively safe area until immediate danger has passed. However, Palestinians have lived under occupation for decades and endeavour to pursue a normal existence. Obstacles to movement and the risks associated with travel through checkpoints or on routes near to settlements create numerous acute and long-term problems for the young. For their part, child protection organisations and international donors have made few concerted efforts to challenge the Israeli authorities over the mobility of Palestinian children and their treatment by soldiers and settlers. As a senior Palestinian children's rights expert related:

> *A lot of resources are being put into developing isolated programming simply because the donors don't challenge Israel on allowing children to travel to different places. The donors will fund the Palestinians to build two or three additional classrooms rather than building one school in a central locality where everyone can come. We'll say "it's cheaper just to buy a bus and to use the bus". And they'll say "no, we can't guarantee that the bus is going to go".*

(3) *Efforts to create social spaces that serve a protective function.* Rather than seeking to transform an existing institutional space into a place of greater protection, some organisations create distinct new spaces. For example, in recent years UNICEF has instigated initiatives such as 'Safe Play Areas' and 'Adolescent Friendly Spaces' working with local authorities and communities to mark out and maintain space where young people have '*opportunities to meet, socialize and play in a protected environment*' (Dolan 2006). Some spaces come into being for a limited time period and specific aims. Most notable amongst these are the numerous summer camps that are held every year involving UN agencies and INGOs together with Palestinian organisations

in a programme of activities intended to '*give young people a safe place to spend their days, a desperately needed chance to have some fun, and an important opportunity to pick up vital life skills to cope with the daily challenges they face*' (UNICEF 2002).

Judging by the large numbers who participate in these various initiatives they are meeting a clear need for social and leisure opportunities otherwise lacking in children's lives. Nevertheless, as with all gatherings of children in a politically unstable setting there are risks entailed. Not only are such spaces vulnerable to violence from Israeli sources, there have also been incidents of attack by extremist elements within Palestinian society affronted by the mixing of girls and boys that typically occurs in many of these projects.

In the oPt, the spatialised approach to child protection – as exemplified by efforts to create distinct and bounded spaces for social interaction – raises an additional and quite specific concern: children's development as members of a larger political collectivity. Through focus groups and interviews we explored with children, parents, and local experts the physical mobility of today's children in comparison to that of older generations when they were young. This involved the use of maps to mark out the places visited on family outings and school trips. In each location of our enquiry the challenges to mobility differed somewhat and were inflected by socio-economic status. Nevertheless, it became apparent that overall children in 2009 had far less possibility to travel even within the West Bank than their parents or grandparents. The options for trips beyond a child's village or town and the immediately surrounding countryside were typically few in number, with Jerusalem and especially Gaza off limits. This has serious implications at a societal level as one government official explained:

> *What protects children is a state, and a state cannot function as a bunch of little cantonettes. The idea that we all share certain commonalities together: that we're all bound by the same laws and principles. Children living in isolated cantonettes never pick up what a state is because they're living in this little area where everything is sort of set apart and they do their own thing.*

Considered in terms of children's right to membership of a national community and to know the heritage of that community, as well as their right of access to family, the obstructions to their movement across the territory of the oPt can be seen as a violation with serious, long-term implications. As suggested by the quote above, the localisation of young people's lives risks not only limiting their cognitive horizons but also fracturing the bonds that might otherwise unite them with a larger collective of people amongst whom they might otherwise reflect upon a collective history and build a common future.

Institutional agendas

In the foregoing discussion, I have described a number of ways in which the spatialised approach to child protection – in the various forms that it takes – may be unsuited to the realities of children's lives in the oPt. Some of the problems inherent to this approach are widely acknowledged: such as the inadequacy of efforts to address intra-household and intra-community tensions in isolation from the pressures associated with life under occupation, or the basic fact that even 'safe space' may be violated at a moment's notice and with impunity. Other limitations of the spatialised approach are noted less often. Nevertheless, there can be little doubt that current protection efforts, in which separation and enclosure are core, have proven ineffective in preventing the systematic violation of Palestinian children's lives and rights. We should therefore seek to understand why spatialisation remains so strongly entrenched in practice.

To some extent the spatialised approach in the oPt continues due to its widespread acceptance within the global child protection field: acceptance that is the product of a tradition of thought about children and within humanitarianism that dates back several decades. However, we need also to consider the institutional purposes of spatialisation that may have little relation to the aim of protection. As with the use of refugee camps for the containment of people who might otherwise prove a threat, so enclosure of children within agency-determined domains may serve a number of important additional functions', which I here mention three by way of illustration.

First, by seeking to work with or support children within distinct spaces agencies circumscribe the arena wherein their own efficacy is estimated. While the protection of children from the violence of the Israeli forces and settlers may never be completely achievable, enclosure narrows the focus of evaluation: within a defined location such as a school, summer camp, or inside their homes improvements to children's lives as a result of agency intervention can be measured.

At the same time, questions about the inability of organisations to influence the conditions of children's lives more broadly, such as to remove the obstacles to their movement or the risks they encounter passing through checkpoints, can be averted. By positively impacting rates of domestic violence or school bullying – worthy aims in themselves – agencies can be seen as effective in enhancing child protection. Thus, it may be fair to suggest that a 'child-friendly space' also serves to define the boundaries within which an agency proves itself to outsiders.

In any setting, humanitarian organisations are likely to be nervous of engagement in matters deemed 'political'; in the oPt, such anxiety is especially pronounced. Here UN agencies and INGOs must operate around the agendas of Western governmental donors who, for the most part have prioritised trade with Israel and appeasement of the pro-Israel lobby over international law and children's rights. At the same time, agencies are faced by the expectations of young people and their caregivers for effective action. In recent years, young Palestinians have been outspoken in expressing concerns and aspirations, attracting international attention: their statements often discussing the violations resulting from occupation and, in some cases, criticising international organisations over their failure to call Israel to account. The spatialisation of child protection activities thus offers a means by which children's voices might be managed. Within bounded, agency-organised spaces, the participation of young people can be 'localised', reducing the risk of their frustration with the role of donor governments in perpetuating occupation becoming a matter of public debate. Concomitantly, children's aspirations can be encouraged towards matters that agencies are able to address, thereby providing the opportunity to demonstrate accountability to intended beneficiaries.

Conclusion

In this brief essay, I have sought to describe and account for the particular dominance of the spatialised approach to child protection efforts as designed at the global level. This was the precursor to discussion of the limitations of this approach when applied to the oPt. My intention in discussing the problematic implications of spatialisation is not simply to draw attention to the need for greater contextualisation of globally-defined approaches – important though this is in an era of increasing standardisation. A deeper aim of this paper has been to consider the obstacles to questioning normative thinking and standardised practice. As I have suggested, the spatialisation of child protection is partly the product of trends of thought within humanitarianism and in relation to children as these have developed through the twentieth century in Europe and North America. In the setting of the oPt, however, we can also see how spatialisation serves the institutional needs of implementing organisations and the agendas of major Western donors. In essence, the pursuit of programming built on the assumption of children's separability from the predations of the occupying forces provides both humanitarian agencies and donors with the opportunity to be seen to be making a contribution to Palestinians' lives while avoiding direct challenge to Israel over its systematic violation of international law. As a consequence the sources of risk to children remain. More worrying still, the evident goal of the Israeli government and settlers of enclosing the Palestinian population within ever-shrinking enclaves is potentially enabled by an approach to child protection that seeks to enclose the lives of the young within 'safe space'. Ironically, such 'safe space' and the inaction that it partially conceals may constitute a considerable threat to young Palestinians in their current lives and for their future.

Acknowledgements

I should like to express my immense gratitude to Claudia Lo Forte for her invaluable work in conducting research that informs this paper and for her collaboration in developing the line of argument expressed herein.

References

Ager, A. (2011) 'Protection of Children in Disaster and War: Assessment and Response', paper presented at Columbia University, New York, USA, 24 March, 2011, available at http://childprotection.wikischolars.columbia.edu/file/view/SLIDES_Session%202_Alastair%20Ager.pdf (retrieved 3 November 2011).

Bronfenbrenner, U. (1979) *The Ecology of Human Development*, Cambridge, MA: Harvard University Press.

CPT & OD (Christian Peacemakers Team and Operation Dove) (2010) 'The dangerous road to education: Palestinian students suffer under settler violence and military negligence', available at www.cpt.org/files/palestine/shh-report_school_patrol_2009-10-the_dangerous_road_to_education.pdf (retrieved 12 January 2011).

Cunningham, H. (1995) *Children and Childhood in Western Society since 1500*, London, UK: Longman.

CZOPP (National Coalition for Children as Zones of Peace and Child Protection) (2011) 'Promoting schools as zones of peace', available at www.czop.org/description.php?url=szop (retrieved 5 November 2011).

Dajani, S. (2005) *Ruling Palestine: A History of the Legally Sanctioned Jewish-Israeli Seizure of Land and Housing in Palestine*, Geneva/Bethlehem: The Centre on Housing Rights and Evictions/BADIL Resource Center.

Dolan, S. (2006) 'Safe-play havens for Palestinian children living in conflict zones', UNICEF-oPt, available at www.unicef.org/infobycountry/oPt_30632.html (retrieved 30 October 2011).

Gillis, J. (2008) 'Epilogue: the islanding of children – reshaping the mythical landscapes of childhood', in M. Gutman and N. de Coninck-Smith (eds.) *Designing Modern Childhoods*, Piscataway, NJ: Rutgers University Press.

Harrell-Bond, B. (1986) *Imposing Aid: Emergency Assistance to Refugees*, Oxford, UK: Oxford University Press.

Hart, J. and C. Lo Forte (2010) 'Protecting Palestinian Children from Political Violence: The Role of the International Community', Refugee Studies Centre Policy Briefing no. 5', Oxford, UK: Refugee Studies Centre.

Jack, G. and O. Gill (2010) 'The role of communities in safeguarding children and young people', *Child Abuse Review* 19 (1): 82–96.

Jenks, C. (1994) 'Child abuse in the postmodern context: an issue of social identity', *Childhood* 2 (3): 111–21.

Lischer, S. K. (2005) *Dangerous Sanctuaries: Refugee Camps, Civil War, and the Dilemmas of Humanitarian Aid*, Ithaca, USA: Cornell University Press.

Malkki, L. (1995) 'Refugees and exile: from "refugee studies" to the national order of things', *Annual Review of Anthropology* 24: 495–523.

Rousseau, J. J. (1969 [1762]) *Émile*, London, UK: Dent.

SCUK (2010) 'What we do in the occupied Palestinian territory', *Briefing Note*, London, UK: Save the Children, UK, available at www.savethechildren.org.uk/sites/default/files/docs/OPT_CB_2009.pdf (retrieved 1 November 2011).

SCUK and SC Sweden (2004) *Living Behind Barriers: Palestinian Children Speak Out*, London, UK: Save the Children UK and Save the Children Sweden.

Tefferi, H. (2007) 'Reconstructing Adolescence after displacement: experience from eastern Africa', *Children & Society* 21 (4): 297–308.

UNICEF (2002) 'News note: UNICEF opens summer camps for Palestinian children', 6 July, available at www.unicef.org/media/media_19505.html (retrieved 3 November 2011).

World Vision International (2005) 'Child friendly spaces (CFS) in crisis situations', available at www.worldvision.org.uk/upload/pdf/What_is_a_Child_Friendly_Space.pdf (retrieved 1 November 2011).

Zelizer, V. (1985) *Pricing the Priceless Child: The Changing Social Value of Children*, New York: Basic Books.

Following the law, but losing the spirit of child protection in Kenya

Elizabeth Cooper

This paper explores how an ostensibly child-centred system can fail to protect children. In some policy arenas, the Kenyan state is recognised as a leader in Africa for the care and protection of children at risk. Yet a case study of children's experiences illuminates how, despite adherence to a legislated framework and series of protocols, the Kenyan state proves unable or unwilling to ensure children's care and protection. The deployment of child-focused discourse and practice through bureaucratic documentation and judicial rulings camouflages (poorly) the state's neglect of children's perspectives and the fundamental risks to children, families, and communities.

Suivre la loi, mais en perdant l'esprit de la protection de l'enfance au Kenya
Cet article traite de la façon dont un système soi-disant centré sur l'enfant peut échouer au moment de protéger les enfants. Dans certaines arènes de politique générale, l'État kenyan est reconnu comme un leader en Afrique en matière de soins et de protection des enfants en situation de risque. Pourtant, une étude de cas sur les expériences des enfants montre comment, malgré l'adhésion à un cadre législatif et à une série de protocoles, l'État kenyan s'avère incapable ou peu disposé à dispenser des soins aux enfants et à assurer leur protection. Le déploiement du discours et des pratiques axés sur les enfants à travers la documentation bureaucratique et les décisions judiciaires camoufle (assez mal) le fait que l'État n'a fait aucun cas des points de vue des enfants et des risques fondamentaux pour les enfants, les familles et les communautés.

Seguindo a lei, mas perdendo o espírito da proteção infantil no Quênia
Este artigo explora como um sistema centrado ostensivamente nas crianças pode falhar na sua proteção. Em algumas arenas de políticas, o Quênia é reconhecido como líder na África para o cuidado e proteção de crianças sob situações de risco. Porém, um estudo de caso sobre experiências de crianças esclarece como, apesar do respeito a uma estrutura jurídica e uma série de protocolos, o Quênia mostra-se incapaz de ou não disposto a garantir cuidado e proteção às crianças. O emprego de um discurso e prática centrados na criança através de uma documentação burocrática e procedimentos judiciais camuflam (mal) a negligência do Estado quanto às perspectivas das crianças e os principais riscos que recaem sobre elas, as famílias e as comunidades.

Cumpliendo con la ley, pero sin respetar el espíritu de la protección de la niñez en Kenia
Este ensayo analiza cómo un sistema aparentemente enfocado en la niñez no resultó eficaz para proteger a los niños y niñas. En algunos círculos de la política, el Estado keniano es reconocido como líder en África por el alto nivel de cuidado y protección que ofrece a la infancia en riesgo. Sin embargo, un estudio de caso sobre varios niños y niñas demuestra cómo el Estado de Kenia

no ha podido o no ha querido asegurar el cuidado y la protección de la niñez, a pesar de ser signatario de varios protocolos y leyes al respecto. El despliegue de acciones y discursos sobre la infancia a través de documentos oficiales y de dictámenes judiciales oculta (sin lograrlo) que el Estado no atiende las necesidades de la niñez ni los principales riesgos que afrontan los niños, niñas, sus familias y sus comunidades.

Introduction

The Kenyan state is recognised in some policy arenas as a leader in Africa for the care and protection of children at risk. The African Child Policy Forum (ACPF) has ranked Kenya first among all African countries for '*its effort to put in place an appropriate legal and policy framework for children*' (ACPF 2008: 1), and Kenya has been evaluated as one of the most child-friendly African governments (sixth among 52 governments). Indeed, with its Children's Act (2001) and associated protocols and personnel for implementation, Kenya seems to have assiduously domesticated the UN Convention on the Rights of the Child (UNCRC 1989) and the African Charter on the Rights and Welfare of the Child (ACRWC 1990).

A study of children's experiences, however, uncovered how disregard for children's perspectives and a lack of accountability for children's care have been enfolded within operations of the Kenyan state's official care and protection system. As a result, some children have directly been put at additional risk, while many children have learned to distrust the care and protection of the Kenyan state.

This paper analyses what leads to such negative experiences and outcomes. It first describes the study and the nature of the state's child protection system. It then provides two case studies that illustrate some of the types of risks faced by children in this system. This leads into an examination of the particular practices and attitudes that influence how the system's procedures are carried out, and in this, how children's care and protection are compromised. The paper's final two sections appraise the system's fundamental flaw – its prioritisation of bureaucratic efficiency over accountability to children – and the actions that might be taken to improve this system.

Methods

The aim of this study was to investigate what happens to children that are found as 'lost' in the western Kenyan city of Kisumu, and taken in under the jurisdiction of the Government of Kenya's Department of Children's Services (the Children's Department). The study was conducted over six months in 2009 by the author (a foreign researcher) and two local research assistants.

Twenty children between the ages of seven and 16 were randomly selected as case study children from the approximately 100 children housed in the Kisumu Children's Remand Centre in January 2009.[1] After gaining the permission of the Children's Department, as well as the Remand Centre's director and social worker, we approached each child to explain the study and to ask for their consent to follow their cases and speak with them regularly about their situations. As part of the informed consent process, we explained to each child that: we were interested in learning about children's personal experiences and therefore there were no right or wrong answers to our questions; we were independent researchers and not working for the

Remand Centre, Children's Department, or any NGO; we would not share an individual's statements with others unless they expressly asked us to do so; we would give each child a pseudonym in our written and oral presentations to ensure confidentiality; and a child could refuse to answer any questions and withdraw from our study at any time without any negative repercussions. We also explained that we could not provide any direct assistance to a child who participated in this study. We repeated these assurances and asked for consent to continue before each instance of asking a child any questions. At the study's outset we planned a referral process in case we learnt that a child was in immediate danger.

Upon gaining consent, we individually or in partners (for translation assistance) conducted multiple (between four and eight) interviews with each child over the six months. Interviews followed a schedule of open-ended questions that asked for children's descriptions of how they came to be at the Remand Centre, their experiences of the Remand Centre, the Children's Department, and children's court, their judgements on where they should live, their experiences of being returned to live with relatives or placed in an orphanage, their worries and hopes, and other related issues. Follow-up interviews included more situation-specific questions, but retained the focus on children's personal experiences, feelings, and judgements. Interviews were conducted in the local Luo language, and held in private settings in the Remand Centre, often before and after a child attended court hearings about his or her case, and, if a child was placed in a relative's or a charity's home, several visits and interviews were conducted with the child in those sites too. This meant that we followed the children to where they moved.

Interviews were also conducted with former and current governmental and non-governmental personnel as well as children's relatives, and participant observation was conducted at the Children's Remand Centre, Children's Department offices, court hearings, police stations, and at the homes to which children were sent during this period.

The Kenyan state and children's care and protection

There is an extensive regime officially in place for the care and protection of children in Kenya. Kenya's Children's Act (2001) closely reflects the wording of the UNCRC (1989), with emphasis on the core principle that all actions concerning children shall be primarily concerned with securing the best interests of the child, which is also meant to allow for children's opinions to be taken into account as may be appropriate. Differing from the UNCRC (1989), but reflecting the content of the ACRWC (1990), Kenya's Children's Act also outlines the duties and responsibilities of a child, thereby emphasising the reciprocal relationship of children with families and communities.

The 'lost child' is an applied category in Kenya's official care and protection regime which refers to a child who is separated from his or her adult guardian(s). The law is set into motion when a child is brought to the government, most commonly by a citizen who either knows or does not know the child (i.e. who 'found' the child as 'lost'), and most commonly via a police station or a local chief's office. As established in the Children's Act, within 24 hours of receipt of a lost child, the police or a Children's Officer must take that lost child before the divisional court so that the child can be registered as a child in need of care and protection and assigned under the mandate of Children's Services. This registration entails the child's remand at a Children's Remand Centre and requires the Children's Department to conduct a Social Inquiry into the child's history and present circumstances to inform a recommendation for the child's placement by the magistrate of the specially-appointed children's court. The Children's Department is then required to bring the child before the children's court every two weeks until the child can be placed by court order in a more permanent living arrangement that secures the child's care and protection. During their time in remand, children do not attend school.

In this study of 'lost children' and the workings of the care and protection regime in the city of Kisumu, I found that for the most part the letter of the law was followed. Children that were brought to police stations were brought before courts within 24 hours and then placed at the Kisumu Children's Remand Centre, where Social Inquiry reports were filled in, and children appeared before the children's court every two weeks until their cases were ruled upon. However, I argue from the basis of that same research that while the *letter of the law* may largely have been followed, there were significant departures from what might be characterised as the *spirit of the law*. The following two case studies of children's experiences illustrate how dangerous such departures can be.

Children at risk

In November 2008, a police officer brought nine-year-old Philomena and her six-year-old brother Simpson to the Children's Remand Centre in Kisumu, western Kenya.[2] The police officer had received the children from a chief in a district 70 kilometres away, to whose office the children had come after their father had threatened to attack them with his machete. According to the children, the father was violent toward them and had been more so since their mother had left them behind with him. From the Remand Centre the children were brought before a magistrate who registered them as children in special need of care and protection by the state and then the children appeared before a children's court every two weeks for four months, while the Children's Department officers were charged with making a recommendation for their permanent placement. Philomena could recall directions to their father's home, but said they did not want to return because their father had told them he did not want them and would kill them if he saw them again. The children did not know where their mother had gone. Eventually, the Children's Officer recommended the two children be repatriated to their father and the court issued this ruling to be carried out.

Here is how Philomena explained what happened after they were returned to their father:

> *My father was still going to drink alcohol and coming back late to beat us up. So we stayed only for two weeks then one day he came back home when he was drunk and told me to go and look for my mother wherever she went with that other man. He said he doesn't want to see me or that he will even kill me so I should just go to find my mother. The next day I went to fetch water and then when I came back home I found him waiting for me. He beat me up and my brother too and he chased us away.*
>
> *It was the second time he was chasing us away so I felt like he did not love us at all. Simpson and I walked all the way to [a market town] when we met a Good Samaritan who asked us where we were going. I told him our story and then he took us to the central police station where we spent a night before going to Winam Court. After going to court, I was again brought back to the children's remand home where I am now.*

We asked where Philomena thought was the best place for her to stay, to which she replied '*I think I should just be here in remand, then after maybe they will find a school for me to go to*'.

Philomena said she and her brother felt lucky to have escaped again and to have found themselves back in the temporary care and protection of the remand centre. What would happen next to them, however, was unclear as the officials charged with their care and protection were at a loss about where to place them since they could not locate the children's mother and there were no spaces available in local charitable children's homes.

In another case, after two months at the Children's Remand Centre, a nine-year-old boy, Okello, was repatriated to the home of his mother and father by court order based on a Children's Department Social Inquiry report. The report simply stated that Okello had wandered off from home with a friend and become lost. However, when we went to visit Okello just three weeks after his repatriation we found he had run away from his parents' home again. We traced him to his aunt's home in Kisumu. He told us that he had run away both times because he had found his mother stealing his antiretroviral drugs (ARVs) that he had been counselled by his doctors and aunt to take every single day to stay healthy. Okello described how his mother did not want anyone to know he was HIV positive and how she had begun taking his drugs for herself. This is why, Okello explained, he had wanted to live with his aunt and not his mother: because his aunt also took her own ARVs and took Okello with her for regular visits to the clinic. Okello wanted to try to stay healthy and judged that his aunt, and not his mother, was his best guardian in this regard. The aunt told us a long story of how Okello's parents had consistently refused to seek medical attention for him, allowing him to become very ill with rashes and tuberculosis (TB), and how she had begged them to let her care for their son since she also had HIV and could empathise with his condition and ensure his care and treatment. Okello and his aunt both seemed relieved to have been reunited, and when we visited again two months later Okello's rashes and TB had been treated.

These cases make it imperative to question what is going on in the operation of the Kenyan state's system of care and protection for children. Both cases demonstrate that despite observance of a relatively strict schedule of protocols – with children's registration as temporary wards of the state, the conduct of Social Inquiry interviews and reports, and the observance of court hearings and orders – children's protection and care were seriously compromised.

Practices and attitudes that marginalise children

This study's key finding was that the state's care and protection system is not organised to be accountable to children's experiences and perspectives. Particular practices and attitudes among child care workers were found to undermine adherence to the principles of pursuing a child's 'best interests' and respecting a child's opinions on matters concerning his or her life. The study found that one of the most crucial lost opportunities in the system is that of the well-conceived but poorly-implemented conduct of Social Inquiry interviews with children, which are discussed next. The specific examples that are cited reflect a common experience of how children's perspectives have been marginalised within the system and how new personnel recruits have been socialised into this organisational culture.

During this study, responsibility for conducting Social Inquiry interviews with children in remand, as well as drafting reports and placement recommendations for the court, was delegated by the Children's Officer to whatever college student was present at the remand centre that week as part of her six month attachment with the Children's Department for her social work diploma (these students were all female in 2009). One day in May 2009 I observed how a new college student was initiated into her work placement at the Remand Centre. Over the morning tea break in the small reception room the Children's Officer, who had stopped in for a visit (his offices were located elsewhere), asked the student, '*How many reports have you completed?*' The student replied, '*Around three,*' and the Children's Officer commanded, '*Today I want them all done; we'll go to court and clear them all!*'

Consequently, over the course of one day, that college student completed 11 different children's Social Inquiry reports with recommendations for the magistrate's orders to be made at the children's court session the next day. I observed some of this work. Instructed by the Children's Officer to fill in answers to all of the questions on the form entitled Social Inquiry, but not trained on how to conduct a social inquiry, the college student sat down with child after child to ask a few questions and fill in the lengthy forms. At one point, the college student asked a sequence of simple questions of a ten-year-old boy concerning his name, his age, his parents' names, his home village and sub-location, the name of his primary school, and the last class he had completed. She then read another Social Inquiry form previously filled out for a different child by a different attachment student, from which she proceeded to copy the answers to fill in the current boy's form as he sat silently before her. In the space for an answer to the question about the child's behaviour, the college student copied out the answer that she had found in the other file: 'deviant'. She proceeded to read the other child's file and to copy notes into the current boy's file under a section which required more narrative detail. As a result of this sleight on the part of the unpaid and untrained college student, who was seeking to satisfy the Children's Officer's demand that she 'clear them all', that boy was labelled as deviant, which did him no favours in the state's management of his care and protection.

On another day when the Children's Officer visited the Remand Centre a different college student on attachment approached him with a girl of approximately 14 years. The college student explained to the Children's Officer, '*I am having a problem with this one*'. The attachment student explained that the girl said she had been living with her grandmother who did not treat her well and who often hit her. The girl had run away to stay with another woman, who was not a relative and who had subsequently taken the girl to a local chief's office. The chief had then brought the girl to the Children's Department. Now, the college student concluded, the girl was refusing to return to live with her grandmother. Instead the girl was saying that she would either live with the other woman or stay at the Remand Centre. The Children's Officer stared into the girl's face and then said in a loud rebuke, '*What do you think this is? A boarding school? This is not a boarding school!*' He then laughed, looking around the room at the other adults, some of whom joined him. As the adults laughed, the girl stood in the room crying silently until the attachment student led her away and wrote a recommendation to the court that the girl be repatriated to her grandmother. When I later asked the student how she found doing interviews with children, she told me:

> *Some are very cheeky. Their stories change. Children often lie. We have to trick them so they give us information. Sometimes we even give them sweets so they think we are friendly.*

Such instances of child protection officers and their delegates taking no account of children's perspectives led to placement recommendations based on poorly-informed and ill-considered judgments. Some of this may be the consequence of a common suspicion that 'children lie'. Therefore, it is worth asking why adults, especially those working as child protection officers, believe this. Based on my other ethnographic research, as well as other existing research concerning adult-child relations in western Kenyan Luo communities, the refrain that 'children lie' is not necessarily consistent with adults' notions of children's moral propensities (Cooper 2011: 150; Grigorenko *et al.* 2001: 369). While there may be some ambivalence toward children's capacities to cunningly protect their own interests, children are not generally thought to be deceptive, especially when questioned by adults to whom they are expected to be deferent (Cooper 2011: 218).

Perhaps it is experience that has made child protection officers distrustful of children's accounts. Remand Centre staff described instances of discovering that certain children had lied about their backgrounds or changed their stories over time. Rather than casting suspicion on all children's stories, it would seem that such experiences should prompt concern for *why* children might choose to lie or change an account. Moreover, our study findings indicated the importance of following the precautionary principle in dilemmas of whether or not to trust a child's account. During our follow up visits to children's court-ordered placements, we had the opportunity to learn whether children's stories had been accurate. Admittedly, some stories had sounded fantastic to us when we first heard them, and we were surprised to learn their validity. For instance, during our visit to 12-year-old Victoria's grandmother's home, where Victoria had been repatriated following a court order, we found it was true that Victoria had been (wrongly) accused of her grandfather's death and chased away by her older sister. The grandmother and sister expressed that they felt extremely remorseful for blaming Victoria and had accepted her back into their care. However, none of Victoria's account of those circumstances had been documented in her Social Inquiry report (although Victoria says she told her interviewer the story) and therefore her placement had been ordered without consideration of what danger she might have faced if the murder accusations were still believed by her relatives.

This study's findings regarding protection officials' failure to adequately solicit, listen to, trust, and investigate children's stories seem to indicate significant neglect in the fulfilment of their professional responsibilities. While it is reasonable to contemplate that the Children's Department and children's court were working to fulfil the ideal that children are best raised by their own families rather than in institutions, this does not explain why the inquiries neglected children's accounts and perspectives. The abbreviated inquiries and sparse documentation of children's cases seem to indicate that expediency took precedence over accountability to children's individual situations and opinions.

The importance of the Social Inquiry reports was evident in the weight they were given by the court in ordering action by state officials for children's care and protection. In all of the 82 children's cases I witnessed in the children's court or for which I reviewed records between January and July 2009, the magistrate made the judgements recommended by the Social Inquiry reports submitted by the Children's Department.

A systematic lack of accountability to children

The experiences uncovered in this study do not seem to be idiosyncratic exceptions to the rule. Indeed, the rule – that is, the law – could be claimed to have been followed in each of these cases. The material from this research demonstrates how the *performance* of an idea of care and protection has been granted such weight that it crowds out the *practice* of care as a *relationship* between people; that is, between adults and children, as well as government agents and citizens.

Insufficient resources

What allows for such systematic lack of care? Certainly, a lack of resources impedes the capacity of the system to work. During this study, there were only two Children's Officers staffed under the District Children's Officer for Kisumu District, as well as three Volunteer Children's Officers, one secretary and one clerk, and the occasional college students on short-term, unpaid attachments for work experience.[3] Budgets were also limited. Children's Officers described how they often could not investigate charitable children's homes, much less private family's homes with children, due to a lack of petrol for vehicles or fare for public transport. The Remand Centre was allocated a quarterly budget of Ksh 20,000 (USD 250) with which to fund the transport of children leaving the Remand Centre for their court-ordered placements. With an average of five children ordered for repatriations weekly (that is, 65 repatriations each quarter), this amount was clearly not sufficient. Our research team learnt this first hand: the cost for one of our research team members to travel by public transport to where just eight of our case study children had been repatriated to live with their relatives was the equivalent of two-thirds of the Remand Centre's quarterly budget for all travel; the cost for one of us to travel with one child to her family's distant location was more than a third of the Remand Centre's quarterly transport budget, which is why this child had remained in remand for over two years.[4] The lack of funding for transport delays children's placements. It also clearly precludes any travel to investigate homes before children's placements by court order as well as follow-up visits to ensure children's care and protection.

Lack of alternatives

There are also limited options for alternative placements for children with relatives who either cannot be traced or who refuse to care for them. The Government of Kenya has recently made it more difficult for charitable children's homes to register and therefore open or continue housing children. The intention of this government policy is to reduce short-term, *ad hoc* solutions (Odul 2009: 29). Without such charitable children's homes for placing children who cannot be returned to their families, however, the only other options are continued accommodation at the over-crowded Remand Centre or institutionalisation at one of the country's rehabilitation schools. Neither the Remand Centre nor the rehabilitation schools offer formal education and therefore children in those institutions will not be enrolled in primary or secondary schools, thus being denied the right to education stipulated in the Children's Act (2001: Article 7).

Without resources to investigate and monitor children's placements, nor ready alternatives to families' care, the efforts of child protection officers are undermined from the outset. The system is too poorly supported to begin to address many of the problems case workers can be expected to confront in seeking to ensure the care and protection of children who have already been rendered vulnerable through separation from their families.

Bureaucratic indifference

Yet lack of resources is not the only significant challenge to children's care and protection by the Kenyan state. An extremely significant problem is how job performance is appraised: in terms of the execution of official procedures, rather than in terms of supporting the best interests of specific children. Some of this is related to the way in which the system is organised to uphold bureaucratic protocol. As other scholars have noted, bureaucracies function to uphold a principle of objectivity and in so doing, can obfuscate the accountability of those who act in the name of the state and the law as well as ward off appeals to individualised considerations or other ethical ideas (Herzfeld 1997; Shore and Wright 1997; Weber 2006 [1922]).

In following the procedural aspects of the law and policy above all other considerations and judgments, state actors – specifically, Children's Officers and the children's court – actively reduce the scope of the law, and diminish the actualisation of its core principles of pursuing the best interests of the child and respecting children's perspectives and participation in decisions that affect their lives. In part, this is attributable to the reductionist orientation of the law. As Clifford Geertz has described, there tends to be a '*skeletonization of fact, the reduction of it to the genre capacities of the law note*' (Geertz 1983: 172). Certainly, there is ample evidence of such 'skeletonization' in the Kenyan state's care and protection regime. As described above, children's histories and perspectives are consistently reduced through the production of Social Inquiry reports designed to simplify placement rulings by the children's court.

Institutionalized neglect

However, the failing is not confined to inadequacies of the state's legal apparatus – in this case, its sparse legal language. Social Inquiry investigations and reports as well as court proceedings and rulings *could* address more. Children's concerns about their relatives' willingness or capacity to care for them could be counted as important information if there was a different idea of the state's responsibility to children being upheld. There could be more questioning of families' roles in children's lives, and more attention to the difficulties that families are facing in caring for children.

The most significant critique of the system, then, issues from how the performance of child protection protocol is used to eschew authentic responsibility for children's well-being. This study uncovers how the system has been organised and practiced to push children back into whatever families can be found and thereby clear children from the public realm and from claims upon public responsibility. The practice of the system seemed more disposed to serving the interests of the state, and that often meant transforming 'lost' children into *hidden* children: many of the problems that originally led girls and boys to lose contact with their families went unresolved, but the children disappeared – or at least, were supposed to disappear – from public and official scrutiny. Paradoxically, the deployment of conspicuously child-focused discourse and practice through bureaucratic documentation and judicial rulings resulted in the tendency to camouflage (poorly and temporarily) the government's neglect of the fundamental risks to children and families.

The most prevalent risks that children in this study felt they faced were poverty and the illness and death of parents and other caregivers. Many children also witnessed physical violence between their parents or guardians, and had physical abuse directed at themselves. Many boys and girls were often forced to leave school before finishing their primary education because of their families' lack of sufficient funds. A few girls in this study described being directed by their relatives to find employment as domestic workers so as to ease the cost of their dependence on their households, and as a result of being abused or unhappy working as 'house girls' they ran away to look for another way to live.

While only a small fraction of children end up in the state's care and protection system, child poverty, orphanhood, and vulnerability are widespread in western Kenya. Nyanza Province is one of the chronically poorest regions of Kenya. An estimated 60 per cent of Nyanza's households live below the poverty line (World Resource Institute *et al.* 2007: 15). As well, one in five adults between the ages of 25 and 49 in Nyanza Province is infected with HIV, while an estimated one of every five children in the province has already lost a mother, a father, or both parents due to HIV/AIDS (Juma *et al.* 2007: i). The combined burdens of poverty and disease leave many families struggling to meet their basic needs. Although the Kenyan government has declared primary education 'free', families incur many costs demanded by local schools and for annual examinations (Cooper 2011: 128). While education is popularly endorsed and esteemed as the surest means for a person to improve his or her economic livelihood, many children are unable to complete primary education, which leaves them and their families questioning what else they might do with their lives (Cooper 2011: 89–91; Onyango-Ouma 2006: 400).

While some important programmes have been implemented by the Kenyan government and its development partners to support children and families, such as budget support for universal primary and secondary education and cash transfers for orphans and vulnerable children, many structural challenges remain for poor families who care for children. It is to these structural challenges that the Kenyan Government and its development partners need to retain their focus to ensure Kenyan children's care and protection.

Improving child protection in Kenya

Nevertheless, it seems that some quite straightforward, and locally appropriate, steps could be taken to improve children's care and protection. First, it is vital that all personnel involved in child protection be carefully trained and sensitised, not just about the rhetoric of child rights and protection, but about their personal and professional responsibilities as the adults charged with particular children's care and protection both during their temporary guardianship and in their placement recommendations. A culture that valorises accountability to children's well-being, rather than bureaucratic efficiency, needs to be distinguished as the strength of the system. Community and home-based outreach initiatives for HIV-affected people are active in Nyanza Province and offer a potential model for a holistic and personalised style of care. However, concerns about the state 'downloading' responsibilities for care to community-based volunteers need to be carefully considered (Brown 2012).

Clearly, additional resources for more thorough and trustworthy social inquiries (including home visits) should be allocated and monitored to ensure that children's placements are being responsibly investigated, implemented, and monitored. With these resources, child protection officers would be better able to perform the responsibilities assigned to them by the laws. Family counselling sessions could be facilitated to help families discuss what other relatives or community members and institutions they might turn to in challenging times. A public and candid reporting system on challenges in the care and protection of vulnerable children might help to facilitate consideration of what broader measures are required to support children, families, and communities.

Other practical steps that seem feasible in protecting children include recruiting the government-appointed chiefs and assistant chiefs for local areas, as well as children's school heads or committees, into child protection strategies. Setting up systems of communication and cooperation between the Children's Department, the Provincial Administrations' network of chiefs, local police officers, local school officials, and the community-based committees established to oversee the new cash transfer programme for orphans and vulnerable children might mean that fewer children's cases need to go uninvestigated and unmonitored. The system's current requirement that a chief bring a child to a police officer who in turn must bring that child to the court for remand means that children's cases take longer to resolve, and without any consultation with local community members. The aim of a more decentralised approach would be to gain from the consultation and cooperation of community-based actors, without dislodging the Children's Department's responsibilities for ensuring children's care and protection.

Conclusion

This study has shown how a system that is supposed to be child-centred can, in practice, further marginalise children and ignore the conditions of their vulnerability. The existence of laws and protocols cannot be trusted as indicators of success in protecting vulnerable children. Rather than being praised for having a legal framework in place, a government's care and protection of children must be judged according to the empirical evidence of children's well-being or vulnerability. This requires looking carefully within the protection system at what is being practiced and how, as well as beyond the protection system at the key challenges that children and their families and wider communities face in trying to live securely and well. To protect and care for the best interests of children, both efforts must be pursued in a spirit of accountability to children.

Acknowledgments

Important gratitude is extended to Winnie Awuor and Maureen Atieno, who shared in the collection of these data. This study was part of a larger ethnographic research project (Cooper 2011), funded in part by the Social Sciences and Humanities Research Council of Canada.

Notes

1. At any one time between January and July 2009, there were approximately 100 children living at the Kisumu Children's Remand Centre, although the Remand Centre was only supposed to house 75 children. Of those children, approximately 15 per cent were classified as 'child offenders' while the rest were children in need of 'care and protection' or 'safe custody'. Approximately two-thirds of all of the children were under the age of 13, while one-third was between the ages of 14 and 16. Approximately 20 per cent of the children in remand were girls.
2. All names of children are pseudonyms.
3. Kisumu District's population was projected to be 576,256 in 2010 (Republic of Kenya 2005: 3). Government services in this district often also serve other populations because Kisumu is a regional urban centre.
4. In the final weeks of this study, our research team faced the ethical dilemma of what to do about the remaining unresolved cases of the children in our study, which included this girl, Philomena and Simpson, and one other boy and girl. We decided to take action to try to secure care arrangements for these children. This involved searching for children's relatives, and if this failed, lobbying managers of charitable children's homes to take in the children. While we were successful in placing this girl back with her family, we could not do so for the other children. Instead, we secured places for three children in local children's homes. Unfortunately, however, we were unable to find a permanent home for one mentally and physically handicapped girl who as of May 2011 was still living in the Remand Centre after three years.

References

ACPF (African Child Policy Forum) (2008) 'How Child-Friendly Are African Governments?', Addis Ababa: African Child Policy Forum, available at www.africanchildinfo.net/africanreport08/ (retrieved 31 December 2011).

Brown, H. (2012 forthcoming) 'Home-based care is not a new thing: unpacking "neo-liberal" health policy in western Kenya', in R. Prince and R. Marsland (eds.) *Regimes of Health, Relations of Care: Anthropological Perspectives on Health and the Governance of Wellbeing in Africa*, Athens, Ohio: Ohio University Press.

Cooper, E. (2011) 'Who cares for orphans? Challenges to kinship and morality in a Luo village in western Kenya', unpublished DPhil dissertation, Institute of Social and Cultural Anthropology, University of Oxford, Oxford, UK.

Geertz, C. (1983) 'Local knowledge: fact and law in comparative perspective', in C. Geertz (ed.) *Local Knowledge: Further Essays in Interpretive Anthropology*, New York: Basic Books.

Grigorenko, E., P. Wenzel Geissler, R. Prince, F. Okatcha, C. Nokes, D. Kenny, D. Bundy, and R. Sternberg (2001) 'The organisation of Luo conceptions of intelligence: a study of implicit theories in a Kenyan village', *International Journal of Behavioural Development* 25 (4): 367–78.

Herzfeld, M. (1992) *The Social Production of Indifference: Exploring the Symbolic Roots of Western Bureaucracy*, Chicago and London: University of Chicago Press.

Juma, M., I. Askew, and A. Ferguson (2007) 'Situation analysis of the sexual and reproductive health and HIV risks and prevention needs of older orphaned and vulnerable children in Nyanza Province, Kenya', Nairobi: Government of Kenya Department of Children's Services, Department for International Development, Population Council and Constella-Futures, available at www.popcouncil.org/pdfs/Kenya_OVC_report.pdf (retrieved 31 December 2011).

Oduol, J. (2009) 'Message from the Secretary for Children's Affairs', *Daily Nation*, 7 May: 29.

Onyango-Ouma, W. (2006) 'School knowledge and its relevance to everyday life in rural western Kenya', *Nordic Journal of African Studies* 15 (3): 393–405.

Republic of Kenya (2005) 'Kisumu District Strategic Plan 2005–2010', Nairobi: Ministry of Planning and National Development, available at www.ncapd-ke.org/UserFiles/File/District%20Strategic%20Plans/Kisumu%20FINAL%20Modified.pdf (retrieved 31 December 2011).

Shore, C. and S. Wright (eds.) (1997) *Anthropology of Policy: Critical Perspectives on Governance and Power*, London and New York: Routledge.

Weber, M. (2006 1922) 'Bureaucracy', in A. Sharma and A. Gupta (eds.) *The Anthropology of the State: A Reader*, Oxford: Blackwell.

World Resources Institute; Department of Resource Surveys and Remote Sensing, Ministry of Environment and Natural Resources, Kenya; Central Bureau of Statistics, Ministry of Planning and National Development, Kenya; and International Livestock Research Institute (2007) 'Nature's Benefits in Kenya, An Atlas of Ecosystems and Human Well-Being', Washington', DC and Nairobi: World Resources Institute.

Children's migration for work in Bangladesh: the policy implications of intra-household relations

Karin Heissler

Drawing on empirical data from ethnographic fieldwork conducted in Bangladesh, this paper examines intra-household relations, and the roles and responsibilities of children in this context. The findings offer several contributions to current debates and approaches in child protection. First, there is a need for greater recognition of intergenerational interdependence both within households among children and their parents, and outside among wider networks in the community. Second, children's work is revealed as having a protective function within these relationships. In the absence of economic improvements and external assistance, communities, which include children, remain a major resource for child protection.

La migration des enfants en quête de travail au Bangladesh : implications sur le plan des politiques publiques des relations au sein des ménages

Sur la base de données empiriques tirées de travaux de terrain ethnographiques menés au Bangladesh, cet article examine les relations au sein des ménages, ainsi que les rôles et les responsabilités des enfants dans ce contexte. Les conclusions proposent plusieurs contributions aux débats et approches actuels concernant la protection de l'enfance. Tout d'abord, il y a la nécessité de mieux reconnaître l'interdépendance entre générations tant au sein des ménages, parmi les enfants et leurs parents, qu'à l'extérieur, parmi les réseaux plus larges de la communauté. Deuxièmement, le travail des enfants est révélé come ayant une fonction protectrice au sein de ces relations. En l'absence d'améliorations économiques et d'assistance externe, les communautés, qui englobent les enfants, continuent de constituer une ressource importante pour la protection de l'enfance.

Migração de crianças para o trabalho em Bangladesh: as implicações da política das relações intrafamiliares

Baseando-se em dados empíricos do trabalho de campo etnográfico conduzido em Bangladesh, este artigo examina as relações intrafamiliares e as tarefas e responsabilidades das crianças em tal contexto. As constatações oferecem várias contribuições para os debates e as abordagens atuais sobre proteção infantil. Primeiramente, há a necessidade de um maior reconhecimento da interdependência intergeracional dentro das famílias entre as crianças e seus pais, e fora das famílias entre redes mais amplas da comunidade. Em segundo lugar, o trabalho das crianças mostrou ter um papel de proteção dentro dessas relações. Na ausência de melhoria econômica e assistência externa, as comunidades, as quais incluem as crianças, permanecem sendo um importante recurso para a proteção infantil.

La migración de la niñez en busca de trabajo en Bangladesh: implicaciones en las relaciones familiares
Basado en datos empíricos de estudios etnográficos realizados sobre el terreno en Bangladesh, este ensayo analiza las relaciones en los hogares y los papeles y las responsabilidades de los niños y niñas en este contexto. Los resultados aportan insumos para los debates y para los métodos actuales de protección de la niñez. En primer lugar, es necesario reconocer más ampliamente la interdependencia generacional que existe tanto al interior del hogar entre padres e hijos, como al exterior, en las redes comunitarias más amplias. En segundo lugar, se demuestra que el trabajo que realizan los niños tiene una función de protección en el seno de estas relaciones. Mientras no haya mejoras económicas o asistencia externa, estas comunidades, con sus niños y niñas, seguirán constituyendo el recurso más importante de protección para la niñez.

Introduction

Recent years have witnessed a keen interest on the part of many policymakers and researchers to learn more about the scale and extent of violence, exploitation, and abuse of children in order to inform child protection programming and policies. Given the hidden nature of many child protection concerns; political and/or cultural sensitivities; denial among some governments of their existence; and stigma associated with many aspects of violence, exploitation, and abuse, data collection of any kind presents both ethical and methodological challenges. Yet while progress has been made in some household surveys to capture numerically – and ethically – more child protection 'issues', the data do not yield information on local perceptions and practices. Thus, there is a risk that the numerical data are analysed without a broader understanding of the context in which they occur.

Interest is also growing in the relevance of local perceptions and practices in developing appropriate, context-specific prevention and response mechanisms for child protection. Nevertheless, the focus of policymakers' attention remains on uncovering numbers of affected children; hence data collection efforts are designed around the scale and extent of issues, and not their underlying dynamics. This focus on the quantitative aspects of child protection risks missing a more comprehensive picture of the realities of girls' and boys' lives most appropriately revealed through qualitative methods, a picture essential to informing suitable programmes and policies. These realities include the factors at individual, household and community level and the impact of ongoing social and economic change, which may lead to or perpetuate practices that can result in harm to children.

Drawing on empirical data from fieldwork conducted with children and adults in rural and urban Bangladesh for over one year between 2006 and 2008 on child labour migration (Heissler 2009), this article examines the nature of intra-household relations and the inter-household factors that shape these interactions. Key themes for exploration include intergenerational relations and the role of work within these dynamics, and the implications of these findings for child protection. Bangladesh provides a good study site because many of the issues raised above play out in research and related policymaking in this setting.

Child protection research in Bangladesh

In Bangladesh, through the influence of large child-focused institutions and the United Nations Convention on the Rights of the Child (CRC), much child protection research focuses on issues of global concern, including children and work, trafficking and sexual exploitation. This is observed in the large number of studies on categories of 'poor' working children in urban centres and, in particular, on groups, such as street children, child domestic workers, and trafficked women and children. Furthermore, many studies of children and work are approached by individual sectors in which the work takes place, instead of taking a perspective that captures more broadly the range of activities in which children engage.

Regarding girls and women, the focus of policymakers in Bangladesh has often been drawn towards a narrow construction of gender, with girls and women viewed as being severely oppressed by patriarchal constraints (Kabeer 2001). This is attributed to the interpretation of the Islamic traditional practice of *purdah*[1] that influences the practices of Hindu and Christian women in the country. High levels of crime, especially increased gender-based violence, domestic violence, rape, and dowry-related conflict, do little to change views about the significant and lifetime disadvantage faced by women and girls in Bangladesh (Sen and Hulme 2006: 114).

The tendency in research, including that supported by policymakers, to focus on exploited and abused children and gender discrimination, draws attention in child protection in Bangladesh to dysfunctional households, social and economic poverty, and discriminatory and harmful practices which adversely impact children, especially girls, across the life cycle. This article argues that there is a need to balance this information with greater understanding of the functioning of households (including children's roles and responsibilities within this institution), the impact of social and economic change on household interactions and decisions, and local practices and perceptions. A more comprehensive and collective understanding of the lives of children, including their interactions with the household and community, is required for better-informed child protection programming and policymaking.

Study sites, research methods, and sample

Ethnographic research was carried out on the factors involved in child labour migration from four villages[2] in Madhupur *upazila* (or sub-district) to other rural and industrial areas in and around Dhaka, the capital of Bangladesh. This focus on children's migration for work necessitated understanding rural childhoods more broadly, including the functioning of households, girls' and boys' roles and responsibilities within this setting, and the impact of social and economic change on this institution. Hence, data were collected on these aspects of social and economic life as well.

Madhupur is situated approximately 160 km northwest of Dhaka. Travel in and out is only by road, primarily by bus. It has sizable Hindu and Christian populations, although Islam is the majority religion; this is also true of the four villages studied.

Agriculture currently constitutes the main source of income in the study sites, in particular, rice and fruit cultivation, livestock, forestry, and fishery. Although the district to which Madhupur belongs is wealthier than many other districts, there is still significant poverty, and many households are no longer able to rely exclusively on their land to meet all their consumption needs. According to census data, almost 64 per cent of households in rural Madhupur own agricultural land (Bangladesh Bureau of Statistics 2007: 39). Official data do not show levels of 'functional landlessness' (defined as less than 0.2 hectares of land) in the villages (Toufique and Turton 2002: 16); however, according to the data collected, approximately 74 per cent of the child migrants in the sample come from such households. Reflecting the national trend, villagers increasingly look beyond the farming sector for sources of income. This has implications for understanding intra-household roles and responsibilities.

As in many other rural parts of the country, work opportunities exist across the four study villages in low-skilled employment, as physical infrastructure (including roads) develops and commercial sectors grow. These jobs are primarily occupied by males, including adolescent boys. In contrast, there are few paid employment opportunities for poor girls and women and most are considered socially unacceptable. They include domestic work in wealthier households and, for older women, road repair or work in a rice mill.

The sample comprised 58 migrant children (35 girls and 23 boys). They were identified using purposive and snowball sampling. The precise age of the participants is unknown because most births are not registered. Their age at first migration varied between approximately seven and 18 years old. In January 2008, their ages ranged between nine and 25 years old. Whereas some of the respondents were no longer migrants when the research began, the majority were between the ages of 14 and 18 years old. Noting that children grow up in the context of their interactions with others, including adults, engagements also regularly took place with another 73 persons, including the children's parents, siblings, friends, peers, and neighbours.

In-depth discussions were conducted with individual children and groups of children, both migrant and non-migrant, and with adults. These were supported by observation and life histories. Collective methods with groups helped set the social and economic context of the research, provided an orientation on the social norms in each village, and revealed key themes. But they also required supplementation with individual methods, which provided perspectives that could not be shared collectively. The findings from each type of method were triangulated. Interviews were carried out in Bangla with the support of a research assistant. I was not fully fluent in Bangla and it was difficult to understand all the village terms and dialects.[3] An assistant helped ensure that I was making myself understood and grasped what villagers were saying. Moreover, the assistant facilitated the individual methods of data collection. Producing two sets of field notes (mine and my assistant's) also helped reduce the likelihood of forgetting or missing what someone had said.

The following section draws on the empirical findings and shows the factors that shape children's roles and responsibilities. We found that those which may lead to protection concerns are primarily shaped by the type of household to which girls and boys belong, and their interactions within this setting.

Households in Madhupur

According to census data, the 'household' is defined as '*persons, either related or unrelated, living together and taking food from the same kitchen*' (Bangladesh Bureau of Statistics 2007: 5). In the study sites, many children live in two-generation households comprising one or more parents and unmarried children, with other family members living nearby, occasionally in the same compound, but usually in another distinct household. A minority of households, often the relatively wealthy ones, contained three generations or more. Typically such a household would consist of parents, unmarried children, grandparents (usually grandmothers were present), and possibly a recently-married son and his wife.

The most usual type of household unit is patrilineal, meaning that line of descent is traced along the male line, and marriage is most commonly 'patrilocal' (after marriage the bride moves to her husband's family home). 'Matrilocal' residence is practised in some households whereby sons-in-law settle with, or in separate households situated near, their wives' parents.

Household composition, birth order, sibling composition, and gender

In addition to the principles of gender and age (Mayall 2003), across a range of socio-economic settings in the majority world (for example, Punch 2001), the factors that affect roles and responsibilities of children within households are the inter-locking dimensions of household composition, birth order, and sibling composition. As explored in an earlier paper (Heissler 2008), these findings are also salient for Bangladesh.

Certain features of household composition suggest outcomes that relate to children's migration for work. According to national data, 15 to 17 per cent of rural households are female-headed, and this figure increases to 25 per cent among the landless (Kabeer 2001: 61–2). In the field sites, many households were headed by females, usually as a result of the remarriage of, divorce by, or death of husbands. These are the least well-off households and are most likely to have children who have migrated for work, which have implications for child protection. For example, 19-year-old Sahara[4] was 12 years old when her father remarried, left the household and stopped supporting her and her siblings and their mother. Not long after he left, Sahara and an older sister migrated for domestic work. Explaining why she had sent Sahara to work, her mother explained,

> *She was very small, but we were very poor. My husband had just left ...*

Roles and responsibilities of children within households are also affected by sibling composition and birth order. Many of the girls in the sample told me that the reason why they had migrated for work was because they did not have older brothers. Not having an older brother creates a problem for households, and especially older sisters, because it puts pressure on girls to support the household, which in turn forces them to challenge hierarchical gendered and generational norms in society. To the question, '*Why are you working here?*' many of the girls who had migrated for work explained it was because they did not have an older brother. For example, 19-year-old Zafrin, who had been working in a garment factory outside Dhaka for the past three years, explained that if she had had an older brother, she would be living in the village and would have had a very different life. Similarly, Beauty, who first migrated for domestic work when she was 16, is now 25 years old. She is the eldest girl of four children and the only child in her family to have worked outside the village. When she was asked why her two younger brothers never worked outside, she explained that they were very young at the time she left for work and '*no older brother [means] it's a sister's responsibility*'.

The domestic life cycle of the household

The composition of households is not static: cyclical changes in size and composition are brought about by births, marriages, and deaths of family members. As with household composition, birth order, sibling composition, and gender, these changes affect the household's means of subsistence.

For example, Beauty (mentioned above) explained that her family used to be very poor, so as the eldest child she left the household to work as a domestic servant in Dhaka. Beauty's family was subsequently able to buy land because of her contributions from her salary. All the income earned (by herself, her father, and later by her brothers) was pooled together. Now that her family owns land and the household is more secure, the roles and responsibilities of each of the children have shifted. For example, Beauty's younger sister has no interest or compelling reason to work outside, and neighbours explain she does not work outside because she is 'rich'. This is a profound contrast to Beauty's experience of the same household many years earlier when they were 'poor'.

Parents' age and health status (aging, sickness, or injury), particularly that of fathers, also impacts the roles and responsibilities of children. It was very common to see widows and married women with elderly and infirm husbands who were physically unable to engage in paid work and who could not financially support the running of the household. This situation is not surprising given the prevailing marriage regime, whereby girls or young women marry men who are, on average, at least five years older than themselves, which leads to a higher probability of widowhood (Amin *et al.* 2006).

This situation was also prevalent in the study sites. Slightly more than 17 per cent of fathers of child labour migrants in the sample were elderly, infirm, or otherwise unable to engage in work that supported the household. For example, Shahin, the elderly father of Taslima (19 years old) and Sukhiron (16 years old), is not able to engage in heavy work. Since Taslima was 14 and Sukhiron was 12, the girls have frequently engaged in periods of domestic work involving migration to other cities or towns to earn money for their household and to cut down on household costs because, when they are working, the costs of their food, clothing and toiletries are covered by their employers.

Although younger than Shahin, another man (Sohel) cannot support his household because of an accident he had many years ago. Moreover, because he has no older son (who would presumably otherwise be working), his two adolescent daughters have to work, and they are employed in a garment factory outside Dhaka.

A close exploration of the household reveals that not all children within poor households are affected in the same way. To help protect children from violence, exploitation, and abuse, and to pursue their best interests, the focus of policymakers and programmers needs to widen beyond the situation of the child to encompass the household and how it functions. Further, there is a need to recognise that roles and responsibilities of members are dynamic and vary according to the individual and domestic life cycle of the household.

Household interdependence

In Madhupur, work is central to childhood, a normal and accepted part of a child's contribution to the household. Much of this work is done within the child's own household. When they are older and physically strong, most boys engage in local paid work such as driving a cycle van or rickshaw, seasonal brick-making, or agricultural contract labour. In contrast, as girls get older, they take on increasingly skilled, physically challenging, and time-consuming work in and for their own households. Hence, children participate in the construction, maintenance, and advancement of the household enterprise. The household should therefore be understood as involving dynamic interdependence between and among members (Alanen 2003). Similar findings have been observed in other contexts: for example, from her findings from ethnographic research undertaken in rural Bolivia, Punch (2002: 131) describes the ties between parents and children as interdependent, and notes the concept is relative, '*worked out in relation to particular needs at particular times in particular contexts*'.

The interdependent relationship may be appropriately envisioned as an intergenerational contract between members, where claims and responsibilities are backed by the rules of wider society (Kabeer 2001: 25). This model allows for consideration of children as social and economic actors, a contrast to some perspectives among researchers and policymakers that consider children as, among other things, '*passive and dependent*' (Lansdown 2005: 10). The nature of interdependence helps explain why many children work and why they are compelled to work to support the household. If parents do not uphold their obligations to their children, this may be seen by children and by neighbours and other members in the community as unfair and unjust; nevertheless, older children will assume responsibility for providing this support. For example,

15-year-old Salmon is from a 'poor' family. His father rarely works and he has a problem with gambling. To pay off his debts, he had to sell off the little land he owned, so now the house and land it rests on belong to Salmon's paternal uncle. Further, Salmon's father had taken a second wife and he was not supporting his first wife, Salmon's mother. At the time the following conversation took place, Salmon was aged 13 and driving a cycle van in his village. He said he would have preferred to be in school; however, he had to drop out after completing Class 3 because his father was not working and Salmon was needed to help the family. Salmon also feels strongly about his duties to support the household to the fullest extent possible. To help fulfil this role he has held a series of jobs in Dhaka and the village ranging from shop assistant, brick maker, and driver of a cycle van. He had to work because:

> *My father has no income and I am thinking about how to take care of my mother and sister. My father doesn't take care [of us] and I am the oldest son so it is my responsibility to take care of them.*

Although neither party has formal means of recourse should the other not uphold their obligations to the 'contract' within the family, it is nevertheless regulated. Neighbours and relatives comment and make judgments if parties to the contract are not meeting their responsibilities. This oversight works both ways: just as Salmon (above) is seen as having responsibilities to support his mother and sister and his father's household, parents are also seen as having responsibilities to their children. For example, on one occasion, on arriving at the home of 18-year-old Vanessa, she was lying on a pile of hay with her back facing her parents' house. Her neighbour, Annika, explained that Vanessa had had a fight with her parents:

> *She [Vanessa] gives her parents clothes and food when she works, but they give her nothing [in return]. At* Eid, *she received no clothes and no hair oil from her parents. We gave new clothes to our children, but Vanessa's father didn't give her anything. All her clothes are old.*[5]

On overhearing this, Vanessa turned and said,

> *I am becoming bad because I am fighting with my parents. I will go [back] to Dhaka. This time I won't come back before five years has passed and I will tell my employers that I have no parents.*

Annika, knowing that Vanessa always gives the income she earns to her parents and brings back gifts, immediately responded, '*you will not do that, you will give all the money to your parents*'. Vanessa shook her head and said very emphatically, '*no, not this time*'.

As this example illustrates, parents are not respected if they fail to take care of their children. This reflects badly on the child and on the status of the family overall. The lack of gifts received at *Eid* illustrates how acutely transgressions between parties to the contract are noticed. It is seen as even more unacceptable when a child in that type of household is known to exceed his or her obligations, for example, by not only remitting income, but also by using savings to purchase necessities for the family, yet still receiving little or nothing in return.

Just as parents may be judged negatively, so too may children if they are not seen to be upholding their responsibilities to the household. For example, 15-year-old Nizam is considered a 'bad' child because he does not listen to his parents and does not give them all the income he earns from working in a brick factory and from painting trucks. He explained that most boys give their full salary to their parents but that he gives them only half because he wants to spend the rest as he likes. Nizam said that his parents and grandparents, who live nearby, are threatening to arrange his marriage because they feel this will force him to be a better son and grandson and to be more responsible. This shows the two-way nature of the intergenerational contract and how parents – even grandparents – can wield more authority over children.

The examples of Salmon, Vanessa, and Nizam provide several points for discussion. Salmon explains his responsibilities as the oldest boy within the household to take care of his mother and sister. This is to be expected in Bangladesh where gender and patriarchy are strong structuring forces. Yet, Vanessa, a middle child and a girl, still feels a strong sense of responsibility to provide for her family. Moreover, she continues to provide even though her efforts are not being reciprocated. As the example of Nizam shows, children are viewed badly if they are not seen to be contributing sufficiently to the household. These examples illustrate the interdependence of household relations shaped by birth order, sibling composition, gender and the domestic lifecycle of this institution. In settings where other social safety nets do not exist, such as Bangladesh, rather than such scenarios being viewed as examples of children's exploitation, dysfunctional households, and the impact of poverty, the mutual expectations and responsibilities between children and parents that they contain may be seen as a protective mechanism.

Inter-household interdependence

Although interdependence was mostly observed at the intra-household level, there was some indication of inter-household interdependence. Although being part of a female-headed household is associated with a propensity for children's work, being the subject of pity may serve as an entry point into a network that facilitates children's employment, and, by extension, the betterment of the household. So, being part of a female-headed household may lead to increased support from others in the neighbourhood who may also be poor, but who may be able to offer some help through the resources they have access to. This is illustrated in the following case study from Hatipur.

Zarina is a mother of four children and is the third wife of a man who usually lives with his second wife. Zarina's eldest daughter Zohura (aged 17) did domestic work for over six years for a family that provided no salary, only food, shelter, and clothing. Zarina was aware that her neighbour Jamila's 18-year-old daughter Arifa had been working for a generous family in Dhaka for many years. In fact, Jamila's second eldest daughter Aklima had worked for that same household previously. The employers had given the sisters money to build up and repair their parents' house, had arranged Aklima's marriage (which included finding a 'good' groom with a good job and paying all the wedding expenses) and had offered to do the same for Arifa. When Zohura was last home visiting her mother, the two of them asked Jamila if she would ask Arifa's employer if he needed extra help. Jamila agreed and it turned out he did need another worker. He subsequently came with Arifa to collect Zohura and brought her back to Dhaka to work in his home. Jamila explained why she had helped them out:

> *Zarina is a poor person, she has no money. Her daughter is getting older and needs money for her marriage. Zarina's husband doesn't give her anything and she is a neighbour.*

Among some policymakers, this example might be viewed as exploitative. Yet, showing the importance of networks and the logic of both intra- and inter-household interdependence, given the absence of social safety nets, women of the same or similar class position may find it prudent to help each other out in the event their luck turns and they eventually need help.

In fact, about eight months later, Jamila's husband passed away and she was seeking financial assistance from close neighbours and kin to help her pay back a loan she had with a micro-credit organisation to meet the family's financial needs, including medical expenses for her husband who had been ill for several years.[6]

Inter-household factors shaping interdependence

The example above suggests how interdependence functions: all members are expected to and, for the most part, feel obliged to contribute to the best of their ability, and in accordance with the factors outlined above. These obligations are not static, but rather are affected by ongoing social and economic change in Bangladesh. Education is one such development that creates new obligations of parents to children.

Over a fairly short period of time, Bangladesh has made great strides towards meeting global commitments in education.[7] State and non-state investments in education have led to a nationwide increase in the number of primary and secondary schools, increased attention to enrolment and retention, and a shrinking female-male gap in levels of achievement, including among the extremely poor (Sen and Hulme 2006: 49). However, despite these accomplishments, among all age groups, many children still remain out of school and there remains a large proportion of over-age children due to late enrolment, low attendance, poor school quality, and repetition (Campaign for Popular Education 2005).

Most of the mothers and fathers in the sample were from 'poor' households. They said it was difficult for them to meet these schooling obligations, which bear financial and emotional costs. For example, starting from the primary level, there are official and unofficial fees and costs that must be paid. Both children and adults were emphatic that parents are obliged to educate their children. As Eliza, a widowed mother of four children (two of whom had done domestic work), explained, it was tough to meet the new expectations around schooling, but '*children say their parents are not good if they don't meet that responsibility*' (see Heissler 2011 for a more detailed discussion of schooling).

Being 'good' children: the system of honour and shame

While children are embedded in households, and their roles and responsibilities are shaped by the intra-household factors outlined above, their thinking and actions are also subject to social regulation as expressed through a system of honour and shame. These norms regulate all aspects of life.

The system of honour and shame is gradually learnt over the course of growing up. Children are expected to conform to these norms, and they also see no benefit in being different. From the time they were babies, the rules of behaviour are constantly being taught to them and reinforced by elders, often their mothers, but also by other girls and boys. Children are chided for breaking the rules, but there is a light-heartedness when the child is particularly young (for example, by light smacking if it involves a toddler or young child). Many elements of honour, such as obeying parents, respecting elders, and not swearing, are therefore picked up by children as part of the process of socialisation rather than being explicitly learnt, although this too does happen.

While some aspects of honour and shame are imparted in the early years of life, with some variation for gender, children are especially instructed about honour and shame around puberty. For girls, the physical transformations that take place around this time mark a key moment for their honour. Often associated with these changes is attention from nearly pubescent, pubescent, or post-pubescent males which they and others consider shameful. The change in girls' clothing style from dresses or trousers to the more conservative *salwar kameez*[8] is an obvious external marker of this transition. Yet, more profoundly, reaching this phase affects their thinking and actions. Though less marked than for girls, pubescent boys too must adjust their behaviour especially towards those of the opposite sex.

Reaching this phase serves as a significant constraint on adolescents' interactions and choices. From puberty, females and males are expected to prioritise and apply these principles to all aspects of their life. At this time, not mixing with members of the opposite sex was highlighted as a key element of honour. Mobility was also revealed as an important theme related to being 'good'. Especially around puberty, not moving far around or outside the village (if you are a girl) and far outside and late at night (if you are a boy) were described as aspects of being 'good'. As learned over the course of fieldwork, these changes were necessitated by the practice of *purdah*, and maintaining it was revealed as another aspect of honour that was especially important for girls when they reached puberty.

Being honourable is fundamental to understanding how interdependence and the intergenerational contract functions and how choices are made. For example, honour was revealed as one's value in one's own eyes, but also the household's place and status in wider society. Men (especially as fathers and husbands) have a critical place in this system because they are seen as the heads of households, so decisions made by other members affect their status in one way or another. Fatema (aged 19), for example, explained:

> *Before marriage it is important to maintain your father's honour because I am known as Samad's daughter and now that I am married it is my husband's honour because I am known as that man's wife ... It is more important to keep your parents' honour than your own honour because a girl's identity is through her father. If I do anything bad everyone will say, 'Samad's daughter is very bad', and it will be my father's shame and that is why I have to protect my father's honour. Now that I am married, I have to protect my husband's honour because now my responsibility is to my husband and not to my father.*

For children in female-headed households, honour was still a vital consideration in regulating behaviour, though the focus of their explanations was on one's own value and that of the household more broadly. As the example of Eliza suggests earlier, the honour system also regulates the behaviour of parents, who may feel a profound sense of inadequacy if they are unable to meet their obligations in the intergenerational contract.

Discussion and conclusions

The empirical evidence yields several findings for discussion. First, not all children in poor households are affected in the same way. Intra-household and intergenerational relations influence why some children work, and migrate for work. In the study sites, the oldest girls in landless or functionally landless households with no boys, much younger brothers, and ill or aging fathers are more likely than their younger siblings to take on extra responsibilities to support the household through migrating for work. Changes that take place over the life course of the individual and household such as the death of a father, father's ill health or remarriage, and aging parents affect relations, with impacts that influence children's work decisions. The system of honour and shame operates at both the intra- and inter-household level, shaping and regulating all aspects of life, including intra- and inter-household relations.

The complexity and dynamism of intergenerational relations have several implications for child protection policy and action. First, development actors must understand how households function and children's roles within them. Linked to this is the need for greater appreciation of the changes that take place over the life course of the individual and household, and the implications of this for children's risk and vulnerability. Appropriately designed and targeted antipoverty measures reaching down to the household level may serve protective functions for girls and boys in the circumstances encountered *if* these considerations are taken on board.

Second, development actors need to appreciate interdependence both within households among children and their parents, and outside households among wider networks in the community. They must further understand the rationale that underpins these relations and shapes decisions, including those about children's work and migration. Communities, including children themselves, constitute a major resource for child protection and household survival. Without fuller appreciation and understanding of the way these communities function, there is a risk that development agents mischaracterise protective relations and actions as exploitative, and that their interventions undermine protective mechanisms. Particularly in contexts with weak social protection schemes, or an absence of economic improvement or external assistance, households and communities serve as the first line of protection for children and families. Policies must therefore find ways of utilising and enhancing these resources rather than hindering them.

Third, much programming, policymaking, and research in child protection is driven by the global advocacy agenda on some of the most egregious violations of children's rights (for example hazardous work, trafficking, and sexual exploitation of children). No doubt it is vitally important to understand and address the situation of children who experience extreme forms of violence and exploitation. Nevertheless, the lives of other poor and vulnerable children also merit attention. Without situating the extreme child protection examples in wider context, child protection work may fail to meet the needs of these girls and boys and their families. Even worse, as mentioned above, it may undermine their self-protective efforts.

Fourth, the objectives of child protection research and data collection efforts must expand beyond a focus on scale and extent of child protection issues to also include the dynamics that underpin them. Further, while child-centred research is growing, providing an important understanding of children's own perspectives on their lives and the world around them, as they grow up in dynamic interaction with others, including adults, it is important that the perspectives of adults are not left out. This also draws attention to situating understanding of childhood – and child protection issues – in broader socio-economic context.

This understanding of the complexity of relations and reflection in interventions is essential if child protection programmes and policies are to be effective and in the interests of children and their families.

Acknowledgements

This study was part of a larger research project, funded in part by the Social Sciences and Humanities Research Council of Canada.

Notes

1. Literally 'curtain' or 'veil'. It refers to the system of seclusion of Muslim women from outsiders. In Bangladesh, to varying degrees, it is practised by women of all religions.
2. I call these villages Amripur, Hatipur, Kuripur, and Rampara.
3. I spoke and understood basic formal Bengali and had taken an intensive language class.
4. As with village names, the identities of respondents have been changed to protect their identity.
5. Vanessa had done domestic work on five occasions previously, starting from when she was nine years old. On her sixth experience working, she switched to garment work. The *Eid* being referred to here is *Eid-ul-Fitre* which falls on the first day following the end of *Ramadan.* On this occasion, family members exchange gifts. It is common to receive new clothes, and children usually get them from their parents.

6. The last phase of fieldwork was completed during this time, and it was not possible to find out if Jamila had received help from Zarina or if Jamila had approached her for assistance.
7. In particular, commitments under the Education for All (EFA) movement, the UN Girls' Education Initiative, and the Millennium Development Goals (goals 2 and 3).
8. Traditional dress worn by females in South Asia. It comprises a long tunic (the *kameez*) worn over baggy trousers (the *salwar).* It is worn with a long scarf that covers the chest area. In Bangladesh, among the middle and upper classes and urban females, the *salwar kameez* is worn by unmarried and married girls and women alike. However, in rural areas, it is uncommon for anyone but an unmarried adolescent girl to wear it.

References

Alanen, L. (2003) 'Childhoods: the generational ordering of social relations', in B. Mayall and H. Zeiher (eds.) *Childhood in Generational Perspective*, London, UK: Institute of Education, University of London.

Amin, S., N. Selim, and N. K. Waiz (2006) *Causes and Consequences of Early Marriage in Bangladesh*, Dhaka, Bangladesh: Population Council.

Bangladesh Bureau of Statistics (2007) *Bangladesh Population Census – 2001 Community Series, Zila: Tangail*, Dhaka, Bangladesh: BBS.

Campaign for Popular Education (CAMPE) (2005) *Education Watch 2005: The State of Secondary Education: Progress and Challenges*, Dhaka, Bangladesh: CAMPE.

Heissler, K. (2008) 'Children's migration for work in Bangladesh: the extra- and intra-household factors that shape "choice" and "decision-making"', *Childhoods Today* 2 (1) www.childhoodstoday.org/download.php?id=15.

Heissler, K. (2009) 'On Being 'Good' in Bangladesh: Child Labour Migration and Choice in Madhupur *Upazila*', unpublished DPhil dissertation, Oxford Department of International Development, University of Oxford, Oxford, UK.

Heissler, K. (2011) '"We are poor people so what is the use of education?" Tensions and contradictions in girls' and boys' transitions from school to work in rural Bangladesh', *The European Journal of Development Research* 23 (5): 729–44.

Kabeer, N. (2001) *The Power to Choose: Bangladeshi Women and Labour Market Decisions in London and Dhaka*, Dhaka, Bangladesh: University Press.

Lansdown, G. (2005) *Innocenti Insight: The Evolving Capacities of the Child*, Florence, Italy: Save the Children Sweden and UNICEF Innocenti Research Centre.

Mayall, B. (2003) 'Generation and gender: childhood studies and feminism', in B. Mayall and H. Zeiher (eds.) *Childhood in Generational Perspective*, London, UK: Institute of Education, University of London.

Punch, S. (2001) 'Household division of labour: generation, gender, age, birth order and sibling composition', *Work, Employment and Society* 15 (4): 803–23.

Punch, S. (2002) 'Youth transitions and interdependent child-adult relations in rural Bolivia', *Journal of Rural Studies* 18: 123–33.

Sen, B. and D. Hulme (eds.) (2006) *The State of the Poorest 2005/2006 Chronic Poverty in Bangladesh: Tales of Ascent, Descent, Marginality and Persistence*, Dhaka, Bangladesh: Bangladesh Institute of Development Studies (BIDS).

Toufique, K. A. and C. Turton (2002) *Hands Not Land: How Livelihoods are Changing in Rural Bangladesh*, Dhaka, Bangladesh: Bangladesh Institute of Development Studies (BIDS).

Child protection and harmful traditional practices: female early marriage and genital modification in Ethiopia

Jo Boyden, Alula Pankhurst, and Yisak Tafere

This article explores divergent perspectives on female early marriage and genital modification in Ethiopia. It contrasts international norms and research evidence with local understandings, the latter focusing on the part these practices play in securing family social heritage, well-being of girls, and their transition to adulthood. The article explains the persistence of these practices in the face of campaigns to eliminate them and questions assumptions behind the international child protection model. It points to unintended adverse consequences of interventions that do not pay sufficient regard to local meanings and social relations, and suggests how policy might be approached differently.

Protection de l'enfance et pratiques traditionnelles préjudiciables : mariage précoce et modification génitale parmi les filles en Éthiopie

Cet article traite des points de vue divergents du mariage précoce et de la modification génitale en Éthiopie. Il met en contraste les normes internationales et les données de recherches avec les points de vue locaux, ces données se concentrant sur le rôle joué par ces pratiques au moment d'assurer l'héritage social des familles, le bien-être des filles et leur transition vers l'âge adulte. Cet article explique la persistance de ces pratiques dans le contexte des campagnes en vue de les éliminer et met en question les suppositions étayant le modèle international de protection de l'enfance. Il met en relief les conséquences négatives imprévues des interventions qui n'accordent pas une attention suffisante aux significations et aux rapports sociaux locaux et suggère d'autres approches possibles des politiques publiques.

Proteção infantil e práticas tradicionais prejudiciais: casamento precoce de mulheres e modificação genital na Etiópia

Este artigo explora perspectivas divergentes sobre o casamento precoce de mulheres e modificação genital na Etiópia. Ele contrasta normas internacionais e evidências de pesquisas com a compreensão local, esta última focalizando o papel que essas práticas desempenham em assegurar a herança social familiar, bem-estar das meninas e sua transição para a maturidade. O artigo explica a persistência dessas práticas em face das campanhas para eliminá-las e questiona os pressupostos que estão por trás do modelo de proteção infantil internacional. São apontadas as consequências adversas não pretendidas das intervenções que não dão atenção suficiente aos significados e às relações sociais locais, sugerindo como a política poderia ser abordada diferentemente.

Protección de la niñez y prácticas tradicionales dañinas: el matrimonio precoz de mujeres y la modificación genital en Etiopía
Este ensayo analiza las distintas opiniones que existen sobre el matrimonio precoz y la modificación genital en Etiopía. Se comparan las normas internacionales y los resultados de investigaciones con la forma en que las personas locales entienden estos fenómenos, enfocando esta última en el papel que juegan estas prácticas para asegurar el posicionamiento social de la familia, el bienestar de las jóvenes y su transición a la edad adulta. Este ensayo explica porqué persisten estas costumbres a pesar de las campañas para eliminarlas y plantea preguntas sobre los supuestos en los que se basa el modelo internacional de protección de la niñez. Advierte sobre las consecuencias adversas no deseadas que resultan de acciones que no toman en cuenta los significados locales y las relaciones sociales, y plantea cómo las políticas sobre el tema podrían abordase de modo diferente.

Introduction

Why is it that the customs of early marriage and female genital modification, which have been shown to endanger the health of girls, persist among some people? In Ethiopia, advocacy campaigns and legal sanctions mean that awareness of the hazards of such practices is widespread. Yet some conscientious parents continue to insist on them, and indeed so do some girls. This article explores the complexity of these practices in the lives of those who adhere to them, and contends that child protection is not always straightforward.

For the purpose of this discussion, we define early marriage as any union between individuals under the age of 18 years, this being the minimum legal age of marriage and the age of majority in Ethiopia.[1] The concentration on girls reflects the fact that early marriages are far more prevalent among females than males in Ethiopia. A survey carried out in 1997 suggested that one-third of girls were married before the age of 15, with significant differences in the extent of early marriage by ethnic group (EGLDAM 2008). In 2005 the median age at first marriage was 16.1 for women, rising to 18 in urban areas and over 21 in the capital Addis Ababa (CSAE 2006). Statistical data confirm ethnographic evidence that early marriage is more prevalent in northern Ethiopia, and that the median age at first marriage is rising in younger age groups.

We take female genital modification (FGM) to denote all procedures involving partial or total removal of the external female genitalia, or other injury to the female genital organs for non-medical reasons (WHO 2010). In 2005, three in four women in Ethiopia had undergone genital modification, the prevalence being lower in urban areas, at just over two-thirds, and very high in the two pastoral regions in the eastern part of the country, with much lower rates among girls than their mothers (CSAE 2006). FGM takes three forms in Ethiopia: cliterodectomy, excision, and infibulation.[2] There are important regional variations in the type of procedure and age of the girls concerned, the extent to which men and women are circumcised simultaneously, whether it is a rite of passage, performed collectively, and who officiates (EGLDAM 2008: 95–6). The EGLDAM survey suggests that in Addis Ababa there is no excision and only cliterodectomy, and in Oromia few excisions and high proportions of cliterodectomy, while in Amhara they report a little more excision (45 per cent) than cliteridectomy (42 per cent). Our data suggest that FGM is no longer practiced in our Tigray site and infibulation, the most invasive form, is not undertaken in any of the sites where we conducted our research.

The findings are based on qualitative research conducted in the context of Young Lives, a mixed-methods, longitudinal, study of childhood poverty which is tracking around 12,000 children in Ethiopia, India (Andhra Pradesh), Peru, and Vietnam. This paper draws on three rounds of data gathered from a sub-sample of 50 boys and girls in five out of the 20 Young Lives sites in Ethiopia, as well as with their peers, caregivers, and community representatives. We conducted focus group discussions with boys and girls and adults in separate sessions, and interviews with a range of community and religious leaders, elders, and service providers. The five sites included one from each of four major regions, plus Addis Ababa, the capital city.

This article begins by reviewing international child protection norms and official policy in Ethiopia. We then explain the cultural and material logics underlying the practices and their importance for the social acceptance of girls transitioning to adulthood. Finally, we draw out implications for policy and practice, and summarise the argument and main messages in the conclusions.

International child protection norms

Female early marriage and genital modification have been widely condemned internationally as both harmful to girls' health and a violation of their rights (Mathur *et al.* 2003; UNFPA 2007; UNICEF 2005; WHO 2010). Ethiopia is a signatory of the UN Convention on the Rights of the Child, which provides for children to be protected against diverse forms of harm, to have a say in decisions affecting them (and stipulates 18 as the age of majority, unless under the law applicable to the child, majority is attained earlier). Ethiopia is also signatory to the African Charter on the Rights and Welfare of the Child, which calls for states to take all appropriate measures to eliminate '*harmful social and cultural practices affecting the welfare, dignity, normal growth and development of the child*', especially those that are '*prejudicial to the health or life of the child*' and/or discriminate on the grounds of '*sex or other status*'.

International concerns about early marriage centre on the patriarchal and gerontocratic values pervading this custom; the relative powerlessness of the girls involved; that marriage under the age of majority is illegal; that early marriages involving bridewealth payments are akin to commoditising females; and that young girls cannot give meaningful informed consent (and indeed are often not even consulted) and are especially vulnerable to manipulation or control (Jensen and Thornton 2003: 14). Other sources of criticism are the known or assumed adverse health consequences, particularly the dangers of early pregnancy and childbirth: maternal mortality and morbidity (Jain and Kurz 2007) (including obstetric fistula, difficult labour), and infant mortality (Mathur *et al.* 2003: 12). Early marriage is also linked with a raised incidence of violence (Jain and Kurz 2007: 8; Jensen and Thornton 2003), abuse of women, and isolation from family and community (Mathur *et al.* 2003; UNICEF 2005). Additionally, early childbearing correlates with lower levels of female education (Jain and Kurz 2007; Jensen and Thornton 2003).

Most criticisms of FGM emphasise the adverse health impacts. There is limited research on health effects of cliterodectomy and excision, though one study suggests that women undergoing excision are more likely to face obstetric complications (Hakim 2001). Infibulation has been associated with severe haemorrhaging, tetanus, sepsis (bacterial infection), urine retention, open genital sores, and injury to surrounding tissue, as well as several longer-term gynaecologic or urinary tract difficulties, which can lead to the death of the mother or child (American Academy of Pediatrics 2010; UNFPA 2007; WHO 2010; WHO Study Group 2006). Also frequently cited are negative effects on girls' and women's mental health (Elnashar and Abdelhady 2007).

Official policy in Ethiopia

The Government of Ethiopia has labelled FGM and early marriage 'harmful traditional practices' and proscribed them in law, the recent draft National Child Policy stating that such customs '*have negative implications for the physical, mental and psychological development of children*'.[3] The ban is supported by a range of preventative measures, comprising advocacy campaigns in schools, through the media, and among local associations, recently complemented by the deployment nationally of female health extension workers with a reproductive health mandate. Remedial interventions include pressure to disclose infringements and conferment of penalties for violations, consisting chiefly of fines and imprisonment (Mekonnen and Aspen 2009). There have been important regional differences in the foci of these efforts, depending largely on the prevalence of the practices locally. Thus, the drive against early marriage has been stronger in the north, particularly in Amhara region, and stronger against FGM in the south, especially in Oromia. In general, the push for abolition has been strongest in urban communities.

Eradication efforts have affected awareness, changed attitudes in some quarters, and may well have contributed to the reduction in incidence in some areas. The increase in the median age of marriage of women aged 25–45 reported in urban areas between 2000 and 2005, and the greater increase in the younger age cohort aged 20–24 in Amhara and Tigray, may be indicative of their impact. Likewise, the much higher decline of FGM in urban areas and among daughters as compared to their mothers in Oromia and the Southern Region where campaigns have been more intense would suggest that interventions may have mattered (CSAE 2001, 2006). Respondents stressed different aspects of interventions as being effective. Thus, in Hawassa, one female leader of an *Iddir* (funeral association) attributed the decline in FGM to effective application of fines and health education in schools, whereas caregivers emphasised the role of awareness-raising activities by a local NGO and work by local government officials with women's associations, youth groups, and school clubs. Girls in Oromia highlighted the effect of school programmes, advocacy by district officials, and educational films.

All of the agents of the state, professionals, and community leaders that we interviewed oppose both practices, as do quite a few caregivers, girls, and boys. Nevertheless, not all calls for reform arise simply from current policy. Narratives of reform also point to changing underlying ideas about the instrumental and symbolic value of FGM and early female marriage in the context of raised aspirations associated with urbanisation, the spread of education, employment opportunities for girls (Woldehanna *et al.* 2011), and an emergent focus on fulfilling children's potential as individuals. In some cases, women's own experiences of problems with early marriage and genital modification have been more pivotal in attitudinal change than government policy. For example, a woman in Hawassa who had endured severe complications during childbirth following genital modification made sure that her own daughters were not circumcised, while a woman from Addis Ababa admitted that she used to defend genital modification as an important local custom, until she witnessed how much a young circumcised girl suffered when giving birth.

Some of the accounts against early marriage reveal a broader disquiet about the power and age differences between men and women in traditional unions, about common assertions that large families have economic advantages, and about the wisdom and justice of women being married off forcibly while young. A woman in Oromia complained about being made to marry in her early teens to a drunkard who squandered most of the bridewealth. A divorced woman in Tigray who was married at 14 and had her first baby at 17 said:

> *I was too young at the time to tolerate all the troubles. I often fell down, and once missed the direction of my home. I was too busy and I could not withstand all challenges and responsibilities. All this embarrassed me and forced me to divorce ... At that time most females were too young for marriage and it was too difficult to withstand the challenges of life like falling from mountains or fetching water. That is why most prefer to divorce ...*

Views about gendered and generational roles, responsibilities, and relations are changing fast, and increased weight is being given to children's – particularly girls' – individual interests, as opposed to their familial duties. For example, concerns that early marriage threatens schooling were voiced by many and a causal link was made in Tigray between the raised age of marriage for girls, improved education outcomes, better employment prospects, and higher salaries. There is also some recognition that young people today are more competent in certain respects than the young in the past, reflected in an emergent idea that allowing young people to play a part in deciding their own fate may, in itself, be protective. Comparing her granddaughter's life with her own at the same age, one woman in Tigray emphasised how empowered young women are today. She contrasted women's past suffering with circumstances of the current generation:

> *... my granddaughter is 14 years old now and if I tell her that she is going to marry, she will reply "marry yourself!" During our time, we were unable to decide on our life but today children have the right to decide ... Children today are very much wise ... Her life will definitely be good because she will be educated and may even marry someone who is educated.*

However, it is intriguing that even with the ban, penalties, changing attitudes, and groundswell of activity nationally, these customs persist in so many parts of Ethiopia; indeed, we uncovered significant resistance to reform. Resistance to the prohibition on FGM is strong even in areas such as Addis Ababa and Oromia, where there have been concerted campaigns against the practice and punishment for infractions is applied. In the next section we outline the logics underpinning FGM and female early marriage in the five sites in order to account for their persistence.

Cultural logics of child protection

Though clear patterns emerged, we found considerable regional, ethnic, generational and individual diversity in beliefs and practices, and yet, rather surprisingly, there was less evidence of major differences based on gender. In the north, particularly in Amhara, early marriage of girls is a predominant feature of the cultural repertoire, whereas genital modification, carried out soon after birth, is of minor importance. In contrast, in the south genital modification is a necessary prelude to marriage and is much celebrated, though early marriage is less common. Also, while there is significant cultural convergence in ideas and practices in Amhara and Tigray, there is considerable variation the other sites, especially the urban ones.

At the same time, Ethiopian society and economy are extremely dynamic and ideas about what is a threat or a benefit to girls are changing fast, such that firmly held beliefs around FGM and female early marriage have led to significant inter-generational disagreement in some families. Thus, some Muslim parents in Addis Ababa had conducted clandestine circumcisions against their daughter's wishes. One girl explained that her mother had arranged for her to be circumcised at an early age, but when the circumciser arrived she managed to escape to a neighbour's house. Another complained that she could not avoid circumcision at age five because so many people in her house that day were determined to go ahead with the procedure. Similarly, a girl in Oromia who was in grade four when her parents arranged her wedding begged her teachers to save her; they tried to change her father's mind. Her grandmother was also against the marriage, and offered her refuge in her house, where she remained for some time. But, her father convinced the teachers that she should return home on the pretext that her mother was sick and eventually forced her to marry. Likewise, girls in Tigray explained that they were beaten by their parents when they reported them to officials for trying to marry them off.

For those who are convinced that FGM and early marriage are intrinsically harmful, it is hard to conceive of the practices as serving a protective function. Nonetheless, local logics embody several protective elements, focusing on the consolidation of family interests in situations of material scarcity and uncertainty, and securing the social place of young women.

Consolidation of family interests

Rates of morbidity and mortality are high throughout the Young Lives sample in Ethiopia: for example, one in five out of the total of 3,000 children have been orphaned of at least one parent. At the same time, most households, especially those in drought-prone rural areas that depend on subsistence rain-fed agriculture, are very poor and highly susceptible to economic shocks (Woldehanna *et al.* 2011). In this context, personal survival is generally contingent upon collective effort, and group solidarity necessarily prevails over individual entitlements, marriage being an important means of consolidating collective and individual interests, as well as a source of social heritage. Controlling the reproductive capacity of females is central to shaping this heritage: arranging a girl's marriage guarantees her social acceptance as an adult, bestows legitimacy on her offspring, and enables her parents to determine whom she marries, when, and with what economic benefits to those involved.

Marrying girls off early, at around puberty, may have particular instrumental value, contributing to continuity in conditions of high maternal and infant mortality (Mathur *et al.* 2003), ensuring young females are provided for materially (Jensen and Thornton 2003: 17) and/or cementing alliances between in-laws as a means of social insurance or social mobility. The economic motivations for arranging for girls to marry early were highlighted by many respondents. In Hawassa, a leader of the Women's Association opposed the plan to marry a schoolgirl off precisely because it was motivated by pecuniary interests:

> *When I asked her why she was forced to marry against her will, she told me that her mother was dead and her aunts had asked her to marry a rich man. They told her that he would relieve their poverty and hers. I was very angry. I told her that I would take her to the legal desk if they persisted ... she is not a means of poverty relief.*

On the other hand, one 16 year-old girl from Tigray, noted that marriage had guaranteed both her economic security and that of her mother.

> *I am happy with my marriage because it was arranged by my parents. It also relieved me of the heavy wage labour, which I have been suffering from. Before, I had no plan of marrying early. But when I stopped school and was engaged in a very tiresome job, I wanted to marry and take rest. During the wedding ceremony our neighbours contributed 600* birr *and my mother some more for the feasts. My in-laws spent much money and grain because they can afford it ... Mother gave my husband on the wedding day 1,300* birr gezmi *[dowry]. The amount was secured from my older sister who works in a factory in a town ... My husband bought me all necessary clothes. After the marriage we are share-cropping the land of my mother. We help my mother in many ways.*

Respondents in Amhara also argued that early marriage extends a woman's reproducing years and hence expands household labour capacity, as well as providing poor or elderly parents with a son-in-law's support. However, a local Kebele (district) official expressed a different view that traditionally early marriage was more common among the rich than the poor because the rich would seek marriage alliances with other wealthy families so as to endow their offspring with land and cattle. While this practice is well recorded in the ethnographic literature, nowadays, with the shortage of land and cattle in Amhara, parents have far less to bestow on newly weds so this rationale for early marriage is weaker.

The type of marriage and whether it involves dowry or bridewealth affects who are the main beneficiaries – the newly weds, or the bride or groom's parents. In Oromia, bridewealth (*gebera*) can be an important incentive for early marriage because it often involves substantial sums of money and several cattle, paid to the bride's family. Indeed, adults in this site considered early arranged marriages as tantamount to exchanging girls for money. However, the need to accumulate bridewealth can also delay marriage, leading many young people to flout tradition by eloping – this being another cause of early marriage.

Turning to the role of the marriage of girls in consolidating a family's social position, adults in Oromia made a major distinction between early arranged marriages and female abductions and elopements. They approved of the decline in abductions, as they '*totally violate all the marriage rights*' of girls and may involve assault, but expressed considerable unease about the rise in elopements, which was perceived as reflecting government promotion of self-determination in matters of marriage, and were especially concerned about the associated decrease in arranged marriages. Adult disquiet about elopement focused on its violation of traditional values around parental authority, young people not having the maturity to act wisely without adult guidance and support, the risk that bridewealth payments might be circumvented and worries about eloping couples lacking a home or a secure income and about the welfare of daughters and their offspring.

Couples frequently organise their elopement in secret and many girls are pressured to elope. Detrimental consequences include the risk of girls and their offspring being abandoned, divorce, and rejection by the girl's family, as well as damage to the girl's and her family's honour, and legal action against the groom and his kin by the bride's family. One woman explained these difficulties:

> *my daughter got married ... She agreed with her husband and married him without our consent ... She was married in a nearby town. She has not come to our home and we have not gone to hers since then .... We are not reconciled yet.*

The girl also confirmed this, saying

> *... my husband paid 2200 birr to my parents as a reconciliatory payment. [However] ... The marriage is also not yet formalised because my husband has to pay the bridewealth which is five cattle.*

Even though the groom's family is expected to compensate the bride's family for violating clan norms, elopement generally results in a lower bridewealth.

In contrast, traditional marriages arranged early are understood as conserving strong intergenerational ties and respect for the authority of adults, while at the same time facilitating the smooth transition of young females into social majority and providing social and physical protection for them and their offspring. Some adults in Oromia did concede that there is a need to adjust customary practice to take account of the couple's views and felt it was wrong to force a girl to marry someone she does not know or like, or who is a lot older. They also argued that girls should not marry before the age of 18, and should have their parents' permission and follow traditional ceremonial procedures. Girls in Oromia also disagreed with elopement, one group arguing that it '*violates the cultural norms of the community*', and should be prohibited. Their reasoning included the adverse impacts on a girl's education, the likelihood of divorce, underage sex and parenthood, and poverty. They stressed the importance in arranged marriages of the ongoing support given to the couple by elders, pointing out that sometimes the groom's parents may even facilitate the girl's education after marriage.

Thus, early female marriage is understood to offer many social and economic advantages to families and kin, enabling adults to shape their social heritage and manage their domestic economy in situations of scarcity. While these norms clearly infringe girls' autonomy and may lock them into detrimental marital relationships at an early age, the logic of consolidating inter-generational ties and familial resources is in itself protective of children in that this ensures an enabling environment for their effective transition into adulthood, the second logic we consider.

Social acceptance for young women

We have argued that marriage is central to the economic and social well-being of women in Ethiopia. In many areas, young women's marriageability depends on their reputation for propriety and competencies in domestic matters, and it is widely believed that genital modification and early marriage uphold these attributes, while also constraining errant sexual behaviour. Hence, the two practices are regarded as highly protective of girls and their offspring, in that they prevent the numerous social and physical risks associated with extramarital sex, spinsterhood, and single parenthood.

In all of the Young Lives sites adults stressed the importance of preventing girls from engaging in extramarital sex, for their own protection and for that of their family and offspring. The risks were cited as including the spread of sexually transmitted diseases, early pregnancy outside marriage, and abandonment by partners, all of which can render girls unmarriageable and lead to their exclusion from family and clan. Specific cases were mentioned to support this concern, as in Addis Ababa where one 13-year-old girl was raped and became pregnant by a local boy, and another died at the same age following an illegal abortion. It was reasoned that by marrying females before or at puberty, they are already spoken for when they reach sexual maturity and hence are less likely to be promiscuous, while genital modification ensures that they are pure and of calm disposition, making them more suited to marriage.

Delaying marriage till 18 based on the couple's choice was claimed by adults in Addis Ababa, Tigray, and Oromia to increase the likelihood of extramarital sex, with all the associated problems already indicated. In the Addis Ababa site, where female sex work is prevalent, raising the age of marriage to 18 was seen to pose many risks for girls, since they '*are becoming "ready" for sex at a much earlier age*'. In Tigray, it was reasoned that families gain status by bringing their daughters up to be chaste and obedient; to have a prestigious marriage a girl must be a virgin and this is far more likely if she marries young. It was also explained that marriage prevents abduction and therefore makes it safer for girls to remain in school. Using the term *ketsige'a*, the importance of ensuring that girls attain an independent life through marriage as young as possible was highlighted and parents expressed eagerness to have grandchildren while they themselves are still fit and well. Respondents associated pregnancy outside wedlock with forced unions and with the possibility of the girl's parents threatening legal redress if the boy refuses.

As with early marriage, by far the greatest imperative behind FGM is moral, and in all four sites where the procedure is performed it is believed to confer purity on girls. The significance of purity in females is particularly salient in southern Ethiopia, as a group of girls from Oromia explained:

> *Traditionally, uncircumcised girls are considered to be unclean which is locally called* lumbutam *– a very harassing kind of term and ... So, girls choose to be circumcised rather than being insulted. Circumcised girls also have a better opportunity of getting a husband ...*

Similarly, uncircumcised women in Hawassa who are not married are described as *koma qerech* ('she remains standing'), a metaphor for 'old maid'.

In some areas, FGM is also thought to influence behaviour. Thus, in Tigray circumcised girls were described as '*humble and obedient*', '*decent*', and '*having good behaviour*'. Similarly, Muslim girls in Addis Ababa reasoned that circumcision prevents girls from developing '*bad behaviour*', such as '*being emotional, out of control, restless, developing sexual need at an early age*', whereas girls who have undergone the procedure are '*cleaner*', and '*calmer*'. FGM was also claimed to prevent clumsiness in the performance of domestic chores, this notion being articulated in a cultural idiom, *yaltegerezech lij qil tisebralech*, meaning '*a girl who is not circumcised will break the drinking gourd*'. Importantly, Wolayta respondents in Hawassa made a direct link between impurity and clumsiness:

> *... if she is not circumcised she will break objects. And if she prepares food who is going to eat it? They consider her as impure.*

In Oromia and Addis Ababa it was girls – not adults – who most favoured genital modification, largely because of peer pressure; this in itself a reflection of the importance girls attach to social acceptance and preserving their reputation in readiness for marriage. While younger girls in Oromia often refuse to be circumcised, teenagers are more likely to succumb to peer pressure. One mother explained:

> *I ... organised the circumcision of my other daughter who is 14 years old. It was done at her request. After she witnessed a girl insulting another who was not circumcised, my daughter came home and asked me to organise her circumcision ... I told her I cannot do that because I could not afford to provide her good food like meat to heal ... But she pushed and told me ... what she wanted was to get circumcised.*

Even though there is no religious endorsement for the practice in Ethiopia and it predates the major monotheistic religions, some respondents regard social imperatives underpinning FGM as being reinforced by their faith. For example, some Muslim respondents in Addis Ababa invoked Sharia doctrine as supporting the practice, despite there being no clear basis for this in Islam; one maintained that '*since it is haram [sinful] to let the girls go uncircumcised, people still cut the genitals of the girls slightly*'. Similarly, a theological rationale within Christianity was outlined by an elderly man in Oromia, who recounted a myth in which genital modification is portrayed as Eve's punishment:

> *Satan lied to Eve saying that Adam had married another woman. She was angry and Satan offered her a herb [medicine] to give to Adam. She tasted the medicine and it went to her abdomen down to the part [the clitoris] that is circumcised now ...*

Some proponents of FGM also perceive the practice to have health benefits. Thus, elders in Oromia maintained that,

> *Circumcision is harmless. It does not bring any health problems for the circumcised girl.*
>
> *Losing blood by circumcision may even wash out some diseases. Thus it is advisable for girls who have certain diseases.*

Similarly, a caregiver in Amhara recounted the case of an uncircumcised woman who was allegedly circumcised by doctors during childbirth – to ensure a safe delivery – showing a strong belief in genital modification being a necessary and protective feature of childbirth.

Thus, in safeguarding family interests and girls' social standing and transition to adulthood, genital modification and early marriage are seen by many as protecting females against social and economic risks such as abandonment, stigma, and destitution. Also, by limiting extramarital sex, insofar as this is feasible, many believe that these practices reduce health risks, even though global standards perceive them as major health concerns. The policy conundrum arising from such evidence is manifest: which criteria do we use to judge what is harmful and what is protective of children? Modern globalised norms emphasise bodily integrity; reproductive, infant, and maternal health; and the rights of girls to be protected from harm and to have a say in matters affecting them, once they reach sufficient maturity. Traditional values are more focused on the collective good and on social acceptance of girls. Both approaches seek to prevent risk and protect girls, but they reflect very different values and understandings and imply the need for very different approaches.

What are the implications for policy and practice?

Given the complex and highly dynamic issues surrounding female early marriage and genital modification, it is not possible to offer simple prescriptions for policy. However, our research evidence does suggest some broad guiding principles.

First, unlike the education and health sectors, child protection policy is about changing values, attitudes, and practices beyond the delivery of professional services; sometimes it entails direct intervention in family life. That these values, attitudes, and practices may have historical and cultural significance and embody positive features means that globalised policies based on *a priori* conceptions of risk and on generic interventions designed without full consideration of local perceptions and contexts are unlikely to become effective as rapidly as intended.

Second, it is not at all obvious that an abolitionist approach backed by punitive measures like imprisonment and fines is suitable for addressing practices of such social significance and intricacy as female early marriage and genital modification. Our evidence is that rather than bringing an end to prohibited practices, punitive measures tend to either transform them, or drive them underground. Girls in Oromia argued that far from eliminating FGM, government intercession has merely created new difficulties. Bullying of uncircumcised girls by those who have undergone FGM has increased and, because parents are reluctant to become involved out of fear punishment, many girls are organising their own clandestine operations, either in defiance of their parents, or to protect them from prosecution. Thus, 35 girls in the Oromia site were circumcised recently and when the police investigated the incident, the girls claimed that they had organised the circumcision without their parents' involvement. In Hawassa, on the other hand, families use their sons' circumcision ceremonies as a decoy for circumcising daughters.

Abolition has had a number of unanticipated deleterious outcomes for girls and the push for change has triggered apprehension about a rise in 'new' childhood risks. Thus, some of the girls in Oromia are resorting to being circumcised at night, rendering the operation potentially far more dangerous. Seemingly, the ban has also lowered the age of FGM in Oromia. Traditionally, girls would be circumcised at around age 15, when they were thought to be ready to marry, but nowadays many girls are anxious to be circumcised well before this age. Hence, by the age of 16 all of the girls in the sub-sample in Oromia had been circumcised, even though three years earlier some were reluctant and one was adamant that she would not agree to it.

Third, identifying a specific practice for elimination, without considering children's wider social and economic circumstances, risks neglecting the most critical problems they face. Enforcing the legal age of marriage and prohibiting FGM in contexts where adolescents are sexually active well before the age of 18 and where there is a shortage of contraception, lack of abortion facilities, and low support for young women who bear children out of wedlock (and then often face divorce and/or become single parents) can result in serious adverse outcomes for young women and their offspring. Moreover, prospects for girls to continue their education at secondary and tertiary levels are currently quite limited in Ethiopia, as are meaningful employment opportunities in many areas for those girls unable to remain at school. This implies that health and other forms of advocacy around female early marriage and genital modification might be more effective if associated with increased access for girls to quality schooling, vocational training, employment, regulation and promotion of appropriate forms of migration, contraception and so on.

Fourth, it is clear that many of the girls and women most affected by current policy do not agree with policymakers about which practices require most urgent elimination (Gruenbaum 2001). It is important to consider what families and girls stand to lose through policy and not just what they might gain, which points to the need for planning policies through consultation with children, their families, communities and other stakeholders. This, in turn, means understanding children's roles, responsibilities, preferences, and aspirations as they grow through childhood and beyond, as well as the values, concerns, and priorities of their families. That said, balancing child protection and autonomy is no easy task, especially since so many girls undergo genital modification and some are betrothed when very young. It is also the case that in some instances the fulfilment of children's right to decide for themselves has led to elopements and genital modification carried out at considerable personal risk against adults' wishes.

Conclusion

Substantial effort has been expended in Ethiopia on measures to eliminate FGM and early female marriage. Together with broader societal changes, such as increased education opportunities for girls, these measures have transformed values and attitudes in many communities, and a significant proportion of our respondents have come to see these practices as harmful. There is a growing sense that girls should be informed about and protected from the hazards of genital modification and early marriage, involved in decisions affecting them, and have the possibility of staying on at school and getting jobs rather than marrying young.

That said, female genital modification and early marriage persist to varying degrees, are valued by many, and are often undertaken with the active consent and even at the insistence of girls. Moreover, the values supporting these practices are not always inconsistent with the well-being of young women; indeed they have traditionally ensured girls' moral and social integration, and thereby their effective transition to adulthood. Many perceive genital modification and early marriage as buffering females, households, and wider kin groups against a range of social and physical risks, the most important protection being that girls find a social place and transition safely to adulthood. This brings into question assumptions about the nature of child protection and poses a dilemma as to whether an approach that emphasises social risks or one that is more concerned with health and individual rights is better for children, and whether or not the two can be reconciled or both can be justified.

In the context of societal and policy transformation, decisions about children's protection involve considerable complexity and uncertainty and must take account of many potentially conflicting factors. For adults, resistance to change often centres on the belief that they know best what risks their daughters confront and that tried and tested customary practice is the most effective means of protection. For girls, it often has more to do with preventing social stigma and rebuke by peers, most likely reflecting their sense of what it means to be an adult woman in their community. Girls make decisions based on their assessments of their chances and parents do what they think best for their children. All must consider the trade-off between early marriage and FGM and other risks, including risks to the family as a whole.

The perceived efficacy of female early marriage and genital modification is manifested in continued resistance to reform, and unintended deleterious consequences in some cases. This suggests that there has been insufficient regard to the socio-cultural and economic context, to the rationale underlying these practices, and to who is designating what is harmful or protective for children. The appropriateness and effectiveness of measures that focus on specific practices in isolation from wider social processes and relations is doubtful; there is also need to consider reproductive health services, and measures to promote women's education and employment.

In the contested and competing repertoires, policy actors need to be wary of the oppositions 'modern' and 'backward', 'rational' and 'ignorant', which permeate the discourse on harmful traditional practices, for such judgments belie the complex moral, social, and practical dilemmas involved in protecting children. When interventions to promote changes in practice do not address legitimate social concerns, nor consider the underlying logics of conventional values and all the potential effects for those involved, they are unlikely to achieve their aims and may bring about resistance and unintended adverse outcomes. This in turn suggests that child protection seen too narrowly, without reference to the meanings of practices for the children's place and their transitions to adulthood within their societies, may in itself become an impediment to children's well-being. Disagreement within families, contestation of state policy, and clandestine actions themselves comprise potential risks to the girls involved.

Notes

1. The 1960 Civil Code of Ethiopia fixed the minimum age of marriage for girls at 15 and for boys at 18. This was changed to 18 for both sexes in the 1987 Constitution which was endorsed by the 1995 Constitution and the 2000 Revised Family Code.
2. Clitoridectomy involves the partial or total removal of the clitoris and, very rarely, only the prepuce. Excision or labiadectomy entails partial or total removal of the clitoris and the labia minora, with or without excision of the labia majora. Infibulation involves removing the clitoris, some or all of the labia minora, and bringing the labia majora together to cover most of the vaginal and urinary opening, often requiring incisions in the labia majora to complete (American Academy of Pediatrics 2010).
3. See the 1995 Federal Constitution of Ethiopia, the 1997 Federal Cultural Policy, the 2005 Criminal Code, the 2000 Revised Family Code, the 2004 National Plan of Action for Children (2003–2010 and Beyond), and the 2011 Draft National Child Policy.

References

American Academy of Pediatrics (2010) 'Policy Statement on the Ritual Genital Cutting of Female Minors', *Pediatrics*, online, available at http://pediatrics.aappublications.org/content/early/2010/04/26/peds.2010-0187 (retrieved 28 February 2012).

CSAE (2001) 'Ethiopia Demographic and Health Survey 2000', Addis Ababa and Calverton, USA: Central Statistical Agency and ORC Macro.

CSAE (2006) 'Ethiopia Demographic and Health Survey 2005', Addis Ababa and Calverton, USA: Central Statistical Agency and ORC Macro.

EGLDAM (2008) *Old Beyond Imaginings: Ethiopia Harmful Traditional Practices* (2nd ed.), Addis Ababa, Ethiopia: Goji Limadawi Dirgitoch Aswogaj Mahiber (Formerly the National Committee for Traditional Practices in Ethiopia – NCTPE).

Elnashar, A. and R. Abdelhady (2007) 'The impact of female genital cutting on health of newly married women', *International Journal of Gynecology and Obstetrics* 97 (3): 238–44.

Gruenbaum, E. (2001) *The Female Circumcision Controversy: An Anthropological Perspective*, Philadelphia, PA: University of Pennsylvania Press.

Hakim, L. Y. (2001) 'Impact of female genital mutilation on maternal and neonatal outcomes during parturition', *East African Medical Journal* 78 (5): 255–8.

Jain, S. and K. Kurz (2007) , 'New Insights on Preventing Child Marriage: A Global Analysis of Factors and Programs', Washington, DC: International Center for Research on Women.

Jensen, R. and R. Thornton (2003) 'Early female marriage in the developing world', *Gender, Development and Marriage* 2 (11): 9–19.

Mathur, S., M. Greene, and A. Malhotra (2003) 'Too Young to Wed: the Lives, Rights and Health of Young Married Girls', Washington, DC, USA: International Center for Research on Women, available at http://www.icrw.org/files/publications/Too-Young-to-Wed-the-Lives-Rights-and-Health-of-Young-Married-Girls.pdf (retrieved 28 February 2012).

Mekonnen, B. M. and H. Aspen (2009) ''Early marriage and the campaign against it in Ethiopia', in S. Ege *et al.* (ed.) *Proceedings of the 16th International Conference of Ethiopian Studies,* 3: 1001–12, available at portal.svt.ntnu.no/sites/ices16/Proceedings/Volume%203/Volume%203%20Complete%20version.pdf (retrieved 28 February 2012).

UNFPA (2007) 'A Holistic Approach to the Abandonment of Female Genital Mutilation/ Cutting', New York, NY, USA: United Nations Population Fund (UNFPA), available at www.unfpa.org/public/pid/407 (retrieved 28 February 2012).

UNICEF (2005) 'Early Marriage, A Harmful Traditional Practice: A Statistical Exploration', New York, NY, USA: United Nations Children's Fund (UNICEF) available at www.unicef.org/publications/files/Early_Marriage_12.lo.pdf (retrieved 28 February 2012).

Woldehanna, T., R. Gudisa, Y. Tafere, and A. Pankhurst (2011) 'Understanding Changes in the Lives of Poor Children: Initial Findings from Ethiopia Round 3 Survey Report', Oxford, UK: Young Lives, available at www.younglives.org.uk/files/country-reports/round-3-survey-report_ethiopia (retrieved 28 February 2012).

WHO (2010) , 'Female Genital Mutilation', World Health Organization Fact Sheet Number 241, available at www.who.int/mediacentre/factsheets/fs241/en/ (retrieved 11 April 2011).

WHO Study Group (2006) 'Female genital mutilation and obstetric outcomes', *The Lancet* 367 (9525): 1835–41.

The authors

Jo Boyden has a PhD in Social Anthropology and has been Director of Young Lives since 2005. Her research has mainly focused on child labour, children and political violence, and childhood poverty – particularly in bringing together academics, practitioners and policymakers to develop effective models and methods for supporting children, their families and their communities in situations of adversity.

Alula Pankhurst (corresponding author) is a social anthropologist with a PhD in Social Anthropology and joined Young Lives as Ethiopia Country Director in November 2009. His role involves representing Young Lives and coordinating the work in Ethiopia. His research interests include community change; household types and cycles; well-being; migration and resettlement; and poverty dynamics. <alulap@gmail.com >

Yisak Tafere is Lead Qualitative Researcher for Young Lives Ethiopia. He has an MA in Social Anthropology and BA in Philosophy from Addis Ababa University and is currently a PhD candidate at the Norwegian Child Research Center, Norwegian University of Science and Technology. He has led four rounds of qualitative research fieldwork in Ethiopia, in collaboration with colleagues in Oxford and Ethiopia. His research interests include cultural constructions of childhood, well-being and transitions; risk, resilience, aspirations, and achievements of children and young people; social protection; and children's work and educational trajectories.

Global priorities against local context: protecting Bhutanese refugee children in Nepal

Rosalind Evans and Rachel Mayer

This paper explores how the United Nations High Commissioner for Refugees' (UNHCR) global priorities and strategies for refugee girls and boys are applied in long-term Bhutanese refugee camps in Nepal. It examines UNHCR's interventions to prevent and respond to child protection issues, including separation from parents and caregivers, and early marriage. These are compared with community perceptions of, and assistance for, children living in difficult circumstances. Young refugees' own research on issues affecting children in the camps offers further insights into how protection is defined and experienced by children living in this context and their suggestions for community and bureaucratic responses.

Priorités mondiales par rapport au contexte local : protection des enfants réfugiés bhoutanais au Népal
Cet article traite de la manière dont les priorités et les stratégies mondiales de l'HCR des Nations Unies relatives aux filles et aux garçons réfugiés sont appliquées dans les camps de réfugiés bhoutanais du Népal. Il examine les interventions de l'HCR afin de prévenir et de répondre aux questions de protection de l'enfance, y compris la séparation des parents et des personnes responsables, ainsi que le mariage précoce. Elles sont comparées à la manière dont la communauté perçoit les enfants qui vivent dans des circonstances difficiles et leur apporte une assistance. Les recherches menées par les jeunes réfugiés eux-mêmes sur les enfants qui vivent dans les camps donnent des informations supplémentaires sur la manière dont la protection est définie et vécue par les enfants vivant dans ce contexte et sur leurs suggestions pour les ripostes communautaires et bureaucratiques.

Prioridades globais em relação ao contexto local: protegendo crianças refugiadas butanesas no Nepal
Este artigo analisa como as prioridades globais da UNHCR e as estratégias para meninas e meninos refugiados são aplicadas nos campos de refugiados butaneses no Nepal. O trabalho examina as intervenções da UNHCR na prevenção e resposta em relação às questões de proteção infantil, inclusive a separação dos pais e cuidadores e casamento precoce. Essas intervenções são comparadas com as percepções da comunidade em relação às crianças vivendo em circunstâncias difíceis e à assistência a elas. A pesquisa realizada pelos próprios jovens refugiados sobre questões que afetam as crianças nos campos oferece ideias adicionais sobre como

a proteção é definida e vivenciada pelas crianças que estão vivendo neste contexto e suas sugestões de respostas comunitárias e burocráticas.

Prioridades globales versus contexto local: la protección de la niñez butanesa refugiada en Nepal
Este ensayo analiza cómo se aplican las prioridades y estrategias globales del ACNUR para la niñez refugiada en los campamentos de refugiados butaneses en Nepal. Examina las acciones del ACNUR para prevenir y enfrentar las problemáticas de protección de la niñez, incluyendo la separación de los padres o tutores legales y el matrimonio precoz. Se comparan estas acciones con las percepciones que tiene la comunidad sobre los niños y niñas en circunstancias difíciles y la asistencia que se les presta. Las investigaciones sobre la infancia que han realizado jóvenes refugiados en los campamentos ofrecen nuevas perspectivas sobre cómo los niños y las niñas que ahí viven definen protección, cómo la viven y qué acciones plantean a la comunidad y al gobierno.

Introduction

This article explores how child protection is defined by international agencies and experienced by young refugees in the context of refugee camps in Nepal. It investigates the application of the United Nations High Commissioner for Refugees' (UNHCR) global priorities and strategies for children in protracted Bhutanese camps by examining their responses to children's separation from families and caregivers and young refugees' early marriages. It explores to the extent which international agencies' universal ideas of child protection overlap with or differ from community perceptions. We examine when and why children and community members choose to rely on self-protection, peer support, and informal networks, and in what situations they access the formal reporting and response mechanisms developed by UNHCR. Throughout the article we refer to 'children' and 'young people' to reflect Bhutanese refugees' multiple and dynamic concepts of childhood and youth, which are not always identical to UN definitions, where chronological age (under or over 18) demarcates childhood from adulthood.

Our research draws on Evans' ethnographic fieldwork since 2005 and Mayer's experience of volunteering in the Bhutanese camps in 2009–10. Data collected during this period has been supplemented by shorter visits to Nepal in 2005, 2009, 2010, and 2011; interviews with UNHCR staff in Geneva in 2005 and 2011; and discussions with refugees in the United States and the United Kingdom between 2008 and 2011.

The article outlines UNHCR's understandings of which children are in need of child protection interventions and why. It presents the standards, procedures, and strategies that the agency employs in conceptualising and responding to child protection concerns around the world, and how they are put into practice in the Bhutanese camps. This is contrasted with young Bhutanese refugees' experiences of separation from parents and caregivers, and early marriage. We analyse the local application of global child protection priorities and strategies through exploring the following questions. How do UNHCR and Bhutanese community members understand childhood, and parent-child or family relations? How do these ideas differ, and why? What are the consequences of these contrasting ways of understanding for young refugees in need of protection?

The article argues that UNHCR, and the liberal states, that fund the organisation's activities and resettle a small number of the world's refugees, adopt a model of childhood popular in the Global North. This entails a belief that young people under 18 should spend their time in education and play, and that even adolescents should not perform 'adult' roles of marriage, child-rearing, work, or combat (UNHCR 1994: ch. 2). These perceptions inform the agency's definition of the types of harm that young people should be protected from. The UNHCR and resettlement countries also favour a particular concept of family in which the role played by biological parents in raising children is privileged over that of extended family members, even in contexts where interdependence amongst wider family networks is the cultural norm. In this article, we question whether UNHCR's universal definition of the risks to young people's well-being pays adequate attention to children's and families subjective interpretations of their situation, including the practical and material challenges they face, and their efforts at self-protection.

Bhutanese refugees in Nepal

The Bhutanese refugees are ethnically Nepali and predominantly Hindu. Their ancestors migrated to southern Bhutan between 1865 and 1930 (Hutt 2003: 24), where their settlement was not resisted by the country's existing inhabitants, who mainly speak Tibeto-Burman languages and practise Buddhism. Yet in the late 1980s the Bhutanese authorities introduced a 'one nation, one people' policy requiring all citizens to wear national dress and follow a Buddhist system of etiquette, *driglamnamzha*. Some Nepali Bhutanese interpreted this as an attack on their cultural identity and protested against the new laws. The Bhutanese Government arrested those who participated in public demonstrations and evicted many southern residents, who settled in Nepal where they have been living in camps since the early 1990s.

The Government of Nepal and UNHCR are responsible for camp management. UNHCR works with implementing partners to provide food, shelter, education, and healthcare. Due to the protracted nature of the refugee situation and consequent donor fatigue, humanitarian provision has been reduced over the last few years. This has resulted in cuts to food rations, fuel, and support for higher education. Refugee families require a cash income to purchase extra vegetables, firewood, and to pay college fees. Some receive a small 'incentive' for working in the camp as teachers and social service providers, while others engage in income-generating activities, such as spinning wool or repairing bicycles. Despite laws prohibiting refugees' employment in Nepal, many Bhutanese work locally as agricultural labourers or teach in private schools.

Since most refugees hoped to return to Bhutan, the Nepal Government held 15 bilateral meetings between 1993 and 2003 with the Bhutanese authorities in attempts to achieve the community's repatriation. However, the failure of these negotiations and of a Joint Verification Process to determine which refugees were entitled to return home led to disagreements over solutions to the protracted situation. Some refugees turned to political violence to enable their return to Bhutan and others pinned their hopes on a future in resettlement countries, such as the USA. For much of the period that this article covers, the future of the refugee community remained uncertain. However, in 2008 a large resettlement programme began in Nepal, and to date almost half the refugees (over 50,000) have departed for countries in the Global North. As a result of the resettlement process new child protection systems have been introduced in UNHCR-Nepal: during the six year period on which this article focuses, an evolution in UNHCR's child protection initiatives has taken place.

UNHCR's global priorities and strategies for refugee children

UNHCR adheres to the standards upheld in the UN Convention on the Rights of the Child (UNCRC), which defines children as those under 18. The organisation's guidelines on working with children demonstrate an understanding of all children as 'vulnerable' and 'dependent', and thus in need of adult protection (UNHCR 1994). The organisation has developed methods based on risk categories including unaccompanied and separated children, children out of school, girl mothers, and girls who experience under-age and forced marriage (see UNHCR 2007) to enable staff to be proactive in identifying children and to '*assess risk factors objectively*' (UNHCR 2007: 2). UNHCR focuses on five global priorities for girls and boys which are: separation from families and caregivers; sexual exploitation, abuse and violence; military recruitment; education; and specific concerns of adolescents (UNHCR 2005: 2). We explore UNHCR's local responses in the Bhutanese camps in Nepal to the first two global priorities – children separated from families and caregivers, and children affected by sexual exploitation, abuse and violence, and specifically early marriage.

UNHCR defines 'separated children' as under 18s who are '*separated from both parents, or from their previous legal or customary caregiver*' and 'unaccompanied children' as those '*children who have been separated from both parents and other relatives and are not being cared for by an adult who, by law or custom, is responsible for doing so*' (UNHCR 2008). UNHCR considers these children to be particularly vulnerable because families '*play a fundamental role in their protection, physical care and emotional well-being*'. Consequently, these children are considered at greater risk of sexual exploitation and abuse, military recruitment, child labour, lack of access to school and basic assistance, and detention, even if they are accompanied by members of their extended family (UNHCR 2005). UNHCR's responses to such children include identification, registration, and documentation, monitoring interim care arrangements, enabling family reunification if it is in the best interests of the child and tailoring durable solutions to their needs (*ibid.*).

In 2004, UNHCR published *Guidelines on Determining the Best Interests of the Child* which specifies UNHCR's policy of considering the best interests of each unaccompanied and separated child under their care prior to resettlement (UNHCR 2008: 31–2). In line with Northern concepts of childhood and family, resettlement countries prefer to resettle children with their biological parents, in most cases reuniting them with their biological mother (Kinch 2008: 102). The process of determining best interests is intended as protection for any child who has been unaccompanied for more than two years and for large decisions affecting their life. However, in practice the process is a '*tool of durable solutions*' (Former UNHCR staff member, 2011) since it is a requirement for all unaccompanied and separated children being resettled. Therefore, panels to determine best interests have developed in countries such as Nepal, Thailand and Kenya where large resettlement operations are being carried out (UNHCR 2008).

UNHCR's second global priority for girls and boys of concern is protection '*from sexual exploitation, abuse and violence as well as HIV/AIDS, teenage pregnancies and harmful traditional practices*' (UNHCR 2005: 2). Along with other '*harmful cultural practices*' which damage '*the health, well-being and development of children*' (UNHCR 2003: 75), early marriage has been categorised by the organisation as an incident of sexual and gender based violence (UNHCR 2003: 18). Early marriage is defined by UNHCR as '*arranged marriage under the age of legal consent*': sexual intercourse in such relationships '*constitutes statutory rape*' and contravenes the (usually female) child's right to be protected from sexual abuse (*ibid.*: 70). Since early marriage often results in girls becoming pregnant as teenagers, it '*threatens the rights of both the child-mother and her baby to life and to full survival and development*' (UNHCR 2003: 70). The agency asserts that '*abuse can be prevented through raising awareness, ensuring improved access to assistance and education as well as safe living conditions and school environments*'. Response mechanisms include provision of '*health care, psycho-social support, measures to ensure the safety of the victim/survivor and legal redress*' (UNHCR 2005: 4).

Local application of global priorities

UNHCR and its implementing partners in Nepal utilise various reporting mechanisms and response strategies to identify and address child protection concerns with the Bhutanese community. General information is gathered from community members through UNHCR's Age, Gender, Diversity Mainstreaming assessments. This process envisages actions targeted towards at-risk groups based on age, gender and other social indicators and involves regular focus group discussions with children, adolescents, adults and elderly people to '*assess protection risks*' and '*mainstream all their views into our programmes*' by asking '*the refugees what solutions they would like to see to their problems*'. For example, children explained in one focus group that buying school uniforms is a problem for some students who cannot afford them, so UNHCR found a donor to cover the cost of uniforms and targeted '*the most vulnerable*' children for this assistance (UNHCR Community Services Officer, 2006).

UNHCR relies on refugees to provide information on individuals in need of protection, whose situation should be reported to Camp Management Committee members. The international agencies also work alongside community-based organisations such as the Bhutanese Refugee Children Forum to raise awareness of child protection matters and encourage community members to report violations of individual rights. The Children Forum engages refugee children who are annually elected to form committees, receive training on children's rights, and monitor protection concerns, such as domestic violence or early marriage in the camps.

> *We have empowered the refugee community to look after the camps themselves. Through groups like the Children Forum, they identify 'vulnerables' and make sure they are cared for.* (UNHCR staff member, January 2007)

UNHCR has engaged Bhutanese refugees in campaigns to raise awareness of the potentially negative consequences of such practices as early marriage and has involved young people in these activities (UNHCR 2003: 78). However, staff members described the challenges UNHCR encountered in acquiring information on these issues since it was filtered through the hierarchical layers of camp management, and because the community has a '*history of hiding things from UNHCR*' to maintain their reputation (Kinch 2008: 61):

> *What we are really interested in is cases where children or girls have been forcibly married, rather than when they have chosen to get married below 18. But what are usually reported to us are love marriages or elopement. It is reported when the community itself does not accept it. We do not usually hear about forced marriages as they are culturally accepted so it is difficult for the girl to come and report them.* (UNHCR Assistant Protection Officer, March 2007)

This insight was supported by Evans' observation of early marriages reported to Children Forum staff in the camp, which often related to the inter-caste nature of young people's marriages, rather than their young age.

When they do receive information about individual children experiencing problems UNHCR investigates. Yet challenges in responding to such cases were reported by a UNHCR Protection Assistant who described the 'legal hurdles' faced in placing children in emergency situations outside the camps, which requires permission from the Nepal Government (March 2007). Other interventions include legal redress in cases of sexual exploitation or abuse, and in a small number of cases UNHCR may use resettlement as a protection tool (UNHCR staff member, January 2009). Additionally, '*until the recent establishment of guidelines for determining best interests, UNHCR had no formal mechanism for interventions in situations where children were experiencing abuse or neglect within their families*' (former UNHCR staff member, March 2011).

In 2006–07, shortly before resettlement processing began, UNHCR and the Nepalese Government organised the first census of the Bhutanese population, which enabled UNHCR to '*learn a lot more about the refugee population*' and to know '*who needs what type of assistance*'. In terms of child protection, '*prior to doing the census UNHCR had no data on unaccompanied and separated children, who we refer to as children at risk*' (UNHCR staff member, January 2009). Staff had assumed that '*the refugee community itself*' was caring for a small number of '*"parentless" or "orphaned" children*' by finding placements to meet their needs (Kinch 2008: 61). However, the census identified 678 unaccompanied and separated children and, since early 2008 when a panel was established, UNHCR has conducted hundreds of determinations of best interests for separated children with a view to deciding whether and with whom they should be resettled (UNHCR staff member, January 2009). The need for such determinations to be conducted for all unaccompanied and separated children prior to resettlement has demanded increased financial and human resources, including the addition of a full-time child protection focal point (UNHCR staff member, April 2010). However, while greater attention to the challenges faced by children in the camps is to be welcomed, we question whether monitoring and response mechanisms that rely on particular categories of vulnerability allow UNHCR to identify and support all children at risk of abuse, neglect, and harm.

Local contexts: young refugees' experiences

The socio-economic context in which children and families live, their livelihood strategies, and the cultural values that shape approaches to child-rearing influence the impact of interventions for child protection. These factors influence the roles parents and children consider necessary and/or appropriate for young people, such as political action, domestic responsibilities, income-generating work, marriage, and/or education. We outline marriage practices and the role of extended family members in caring for children amongst the Bhutanese community, drawing on research conducted by 42 young refugees (aged 12–17 years) facilitated in 2007–08 by Evans and staff of the Bhutanese Refugee Children Forum. This organisation supported a group of young researchers in each camp, who had received training in research methods and ethics. The research topics were chosen by the children themselves and in part reflected concerns of the Children Forum, including early marriage, elopement, children living without their parents, children in single-parent families, child labour, children's domestic work, domestic violence, and school drop-outs. This initiative enhanced our understanding of the ways both refugee children and adults articulate child protection and propose solutions, many of which differ from UNHCR's universal definitions and responses.

Marriage and family practices

Within the Bhutanese community, in line with Hindu religious beliefs and social norms, many refugee adults married relatively young, usually between the ages of six and 18 years. Refugee daughters-in-law are expected to move to their husband's family's home, where they often '*have to engage in household work from morning to night*' (refugee woman, October 2006). Polygamy is legal in Bhutan and although officially illegal in Nepal, the practice has continued amongst the refugees. Inter-caste marriage appears to be a relatively recent phenomenon amongst the refugee community, as is the idea that, in line with Nepali law and the international norms promoted by UNHCR, young people should not marry below the age of 18 years.

Most Bhutanese refugees live alongside their extended families where, in addition to their biological parents, other family members also play an important role in raising children. As was the practice in Bhutan, refugee children whose parents have died, remarried, or gone missing usually stay with their patrilineal relatives (Kinch 2008). It is often difficult for divorced or widowed mothers to remain in the same household as their children and to maintain a close relationship with them. When a husband and wife divorce, the mother is expected to leave her husband's household and often has to leave her children behind unless they are infants. Further, if a mother remarries after the death of, or separation from, her husband, she may not be able to bring her children from a previous marriage particularly if they are of a different caste to the new husband (Kinch 2008: 76–7). For example, Samhita explained:

> *My Mum lives in another camp and is married to someone else ... My Mum and Dad divorced when I was two years old so I live with my uncle and grandmother and they love me.* (14-year-old girl, April 2011)

Children's separation from their biological parents following remarriage is a common and accepted practice amongst the Bhutanese refugee community and in some cases there is no close bond between the biological parent and child, whose attachment is likely to be to other members of their extended family (Kinch 2008: 84).

Both the young researchers and a study conducted by a UNHCR Intern, Tara Kinch, found that the largest number of unaccompanied and separated children were those who had been separated from their biological parents due to their mothers remarrying following the death of, or abandonment by, their father. The majority of those identified by UNHCR following a rapid assessment in 2007 were aged 12–18 years and only 1.3 per cent were under four years old. Most '*had at least one parent residing in one of the seven camps*', but were '*cared for by extended family members*' (Kinch 2008: 63), suggesting that '*the cultural implications of remarriage on children from previous marriages are the most prevalent cause of parent-child separation in the camps*' (*ibid.*: 76).

Children without parental support

The young refugees chose to conduct research on both 'children living without their parents' and 'single parent children'. Children without their parents were defined '*as those children under 18 who are living without their father and mother at their hut, whether this is because their parents are dead, eloped, or missing*' (Child Research Report 2 2008: 31). Although the Nepali words *Tuhura* (male) or *Tuhuri* (female) directly translates to 'orphan', it also refers to children '*who have only one parent living with them*' (*ibid.*: 23) and has negative connotations as one young refugee explained: '*some people treat us in a good way, but others call us bad words like* Tuhura-Tuhuri' (*ibid.*: 37).

The concept of single parent children adopted by the young researchers is instructive on how young refugees perceive parents' roles and responsibilities: '*we have defined single parent children as those children who are under the care of only one parent because of the other parent's death, elopement, loss, psychosis or separation from their children*' (Child Research Report 2: 23). Even in situations when both parents are physically present, the young researchers added a 'parent at risk' category, when '*a boy or girl is considered to be a single parent child*' if '*either one of his or her parents is mentally ill and is unable to identify or care for his or her son, daughter, friends, relatives and neighbours*' (*ibid.*: 24). The researchers in this group asked other young people, community members, and teachers to give their opinions on who 'single parent children' are. As well as children living with a 'parent at risk', one refugee teacher placed children living with elderly parents in this category (*ibid.*). Research participants were also asked who they consider to be a parent. Their responses reveal that '*simply giving birth*' to a child is not enough, since a parent is '*someone who creates an environment to support the best interests of the child*' (teacher, *ibid.* 26), '*someone who can look after the health of the child*' (student, ibid.), and '*someone who looks after the all round development of the children*' (parent, *ibid*: 26). Additionally, as one nine-year-old child explained only '*a loving father or mother is called a parent*' (*ibid*: 26) These definitions, and the fact that children whose parents are elderly or ill are regarded as living in 'single-parent' families, suggests that being a parent in this community entails more than simply being the biological mother or father of a child. The role of a parent involves loving a child and demonstrating the capacity to provide for them and support their best interests.

In keeping with this understanding of parents' and/or caregivers' duties, many similar problems were shared with the young researchers by children living without their parents and children in single-parent families. These children felt that they had fewer friends than other children and experienced social discrimination: '*we are not provided proper love and respect by responsible persons in society*' (children without parents, *ibid.*: 34). Some described their sense of insecurity because '*we are all children at home and there is no responsible person to take care of us*' (child without parents, *ibid.* 34), and girls feared sexual exploitation or attacks in the absence of a dependable adult in their hut (single parent child, *ibid.*: 26). Other challenges include time-management problems such as being unable to focus on studies, in addition to completing domestic tasks: '*I generally wake up at five a.m. and start doing household work and because of this I don't get time to read and do my homework*' (child without parent, Child Research Report 2 2008: 34). Young people with these additional responsibilities at home, especially older siblings, often reported being late for school or having to miss school: '*generally, on Wednesdays I don't go to school as there is ration distribution for us*' (child without parent, *ibid.*: 35), when a member of the household has to collect food items provided on a fortnightly basis. This often resulted in children getting into trouble: '*when we are late to school, our teacher scolds us and gives us punishment*' (child without parent, *ibid.* 35).

A further difficulty suffered by some children living without parents and in single-parent families is economic hardship, including '*sitting without food*' when the ration has run out or after being compelled to sell food to '*fulfil our needs*' for firewood, clothing or school fees to study outside the camp after Class 10 (children without parents, Child Research Report 2 2008). These children and families also mentioned that other community members will not loan them money in times of need as they do not believe they will be able to pay it back (*ibid.*: 25, 35). As mentioned earlier, all households need a cash income to supplement food rations and pay for extra items including school uniforms, shoes, and toiletries. In addition to children living without their parents or in single-parent families, those living with ill, alcoholic, elderly, or disabled parents or with step-parents also reported spending much of their time on domestic and income-generating tasks. Such responsibilities were often greater for eldest siblings: '*I have to work from morning to evening to care for my small brothers and sisters because both my mother and father are drunkards*' (15-year-old boy, *ibid.*: 54). Children undertake various income-generating activities, including spinning wool, collecting stones in the river bed, construction work, and bus conducting.

Although some of these children face time management and economic difficulties, teachers and community leaders explained that they often develop competencies ahead of others. Children from single-parent families were observed to be more responsible, studious, and helpful than their peers and often did better at school (Child Research Report 2 2008: 27). Their parents appreciated the support their children provided, which they compared positively to other children's behaviour. Some adults explained that children from single-parent families were less likely than others to engage in anti-social activities, since they spend their time working. While some children reported discriminatory treatment due to their 'orphan' status, others received encouragement and assistance from community members: '*In some cases my household works have already been done by the neighbours*' (child without parent, *ibid.*: 37).

In line with the understanding of parental roles implied in the young people's research, some children both of whose parents were present but unable to fulfil their practical and caring responsibilities, experienced many similar challenges as those living without their parents. For example, in one family where both the parents of three small children abused alcohol, the children suffered from lack of shelter and food as '*the walls of their hut fell down because their parents didn't repair it and they burnt the hut for firewood as they had no kerosene or briquettes*' (refugee neighbour, November 2006). A neighbour reported '*the father barters almost all their rations for alcohol so the kids are often hungry, but when the wife tries to keep the rations the father threatens to kill her*'. Despite the neighbours informing UNHCR about this case, it was other refugees living nearby who ensured the children received food and were not sleeping outside. The recent resettlement-driven focus on unaccompanied and separated children in the camps may draw attention away from those children living with both their biological parents who are unable or unwilling to care for them.

Early marriage

The young researchers classified '*marriage between people below the age of 18 years*' as '*child marriage or early marriage*'. This occurs '*when boys and girls are forced by their families to get married and also when they make the decision themselves and have a love marriage*' (Child Research Report 1 2007: 17). Another group of researchers investigated young people's experiences of elopement, defined as '*girls marrying secretly with their boyfriend, or vice versa, without the permission of their parents*' (Child Research Report 2 2008: 40). The young researchers noted that '*previously child marriage was not considered a problem in our society as it was a normal part of our culture and tradition, but now we have realised the negative effects of early marriage on children*' (Child Research Report 1 2007: 21). However, '*in Nepal, child marriage is illegal so people do it in a hidden way in our camp*' and '*this makes it difficult to find out how many people experience early marriage as it is often not reported*' (*ibid.*). Further, although refugee children are encouraged to '*bring problems to the Children Forum, like early marriage*' so that they can be reported to UNHCR (Children Forum member, March 2005), since early marriage remains widely accepted amongst the refugee community, this can bring the Children Forum into conflict with Camp Management Committees:

> *The agencies expect children and the Children's Facilitator to report cases of early marriage and domestic violence. However, some people in the Camp Management Committee and/or other community members prefer to settle such cases within the community itself and are unhappy when the Children's Forum brings these cases to the attention of people outside the refugee camps.* (Child Research Report 1 2007: 27)

This may explain why there were only 14 cases of early marriage reported in 2005, and nine in 2006 for all seven camps (Child Research Report 1 2007: 17), despite the fact that doctors responsible for providing health services to the refugees observed a high level of teenage pregnancy amongst the community, suggesting that many early marriages remain unreported (*ibid.*: 21).

The young people's research identified many reasons why young people marry early. Some experienced pressure from their parents or other relatives and referred to the '*cultural traditions of our society*', which underpin such social expectations. For example, one woman who married when she was 13 reported, '*My family believed that girls have to marry before menstruation to be able to go to heaven*'. (Child Research Report 1 2007: 18) In addition to older relative's cultural values, some girls living without their parents reported being compelled to marry young. A female participant disclosed that her caregiver, an extended family member, suggested she marry early so that she would move to her husband's household (Kinch 2008: 99). When young people eloped rather than married in accordance with their parents' or relatives wishes, this was often a result of relationships and sexual encounters. Young women's reasons for marrying below the age of 18 included love for their boyfriend and teenage pregnancy: '*I became pregnant before marriage and I had a one and half month old baby in my womb when I eloped!*' (Child Research Report 2 2008: 44).

The young refugees' research reports also suggest that some girls decided to marry as a means of self-protection from situations of domestic violence or marriage to an undesirable partner. For example, one girl who was 16 when she married explained,

> *I was beaten by my maternal family and I had to bear tension all the time. When they saw this, my friends suggested to me that I get married and I thought I would get rid of such tension.* (Child Research Report 1 2007: 18)

Another young woman, Sunia, who '*fell deeply in love with a boy when she was studying in class 8*', eloped with her boyfriend '*after the parents of another boy came to her house from Bhutan and asked her to marry their son*' even though '*she felt the boy from Bhutan did not understand her*' (Child Research Report 2 2008: 44). Since UNHCR's early marriage reporting mechanisms are often utilised by parents or older relatives who disagree with a relationship, these do not always work to protect children, especially in situations where young people choose early marriage as a self-protection strategy in response to domestic violence or pressure to marry someone of their parents' choice.

For boys, the decision to marry early (either made independently or as a result of a request from older family members) was often related to their family's needs, and occurred in situations where there were no female relatives able to cook and do domestic work:

> *There is no one in my home who can work, as my mother is frequently sick and I am not able to read much. I needed support at home so I asked a girl to elope.* (Child Research Report 2 2008: 44)

In many situations boys interpreted the decisions to marry young positively and as a means of fulfilling their duty to contribute to their family's well-being.

Conflicting concepts of childhood and child protection

The approaches of UNHCR and the Bhutanese community to child protection demonstrate variations in understandings of appropriate roles for young people, in concepts of family, and in parent-child relationships. Such differences influence how child protection is perceived by international agencies and Bhutanese community members, and the solutions they initiate. These insights are relevant not only to Bhutanese refugees, most of whom will now be adjusting to life in the Global North, but to other refugee groups where UNHCR is responsible for responding to the needs of child protection.

UNHCR's policies and practices demonstrate the organisation's adherence to the concept of childhood prevalent in the Global North, which has become globally influential partly as a result of a particular interpretation of the UNCRC. This model developed under specific historical conditions in Europe (Cunningham 1998), and views childhood as a time for education and play, and children as morally and physically vulnerable and in need of protection from adult arenas of work, sex, and politics. UNHCR and the liberal resettlement states hold an ideal of family in which the role of biological parents in raising children is given preference over that of extended family members, and young people's sense of duty towards their family is unacknowledged. This model of childhood is often assumed to be universal and natural, which detracts attention from questions of cultural and contextual differences in experiences of childhood, and results in the use of predetermined ideas of the risks young refugees face, and the situations or practices that they should be protected from.

In contrast, as the young researchers' findings indicate, Bhutanese refugee children live in a society in which interdependence amongst extended family members is the norm and being cared for by relatives other than parents is culturally accepted. These children display agency in pursuing self-protection strategies and draw on support from peers and extended family members. However, this model of childhood and family does not conform to the dominant views of children's vulnerability and dependence suggested in UNHCR's guidelines and risk categories.

UNHCR's limited resources for child protection currently focus on determination of best interests for unaccompanied and separated Bhutanese children in the context of a large resettlement operation. This focus on unaccompanied and separated children fails to acknowledge the many different family situations and relationships in which young people may be exposed to harm. However, the difficulties shared by young refugees living without their parents or in single-parent families pre-date the resettlement process. They are related to everyday social discrimination, practical struggles in managing to study and complete domestic chores, and the economic challenges experienced in families where there are no adults able to engage in income-generation work. The young refugees' research demonstrates a much broader definition of parental roles, in which caring and providing for children is more important than biological parenthood. This understanding of parenthood allows recognition of the varied contexts in which young people may need additional support to enable their physical, emotional, and social well-being in the absence of their parents or extended family's ability to provide adequate love, shelter, and food.

Young people and their families may interpret risks and seek protection in ways that differ from the universal concepts of the UNCRC. Early marriage is an example of a practice that can be both a protection strategy and a protection concern (Talviste 2009: 8).The differences in cultural values, concepts of family, and ideas of parents' and young people's responsibilities begs the question of who decides what is in the best interests of the child, and on the basis of what values. However, UNHCR's universal categorisation of the risks to young people's well-being leaves little room for children's and family's subjective interpretations of their situation and their own efforts at self-protection. In contrast, the young researchers' findings recognise refugee children's agency in pursuing self-protection strategies and drawing on support from peers and extended family members, which may not conform to the dominant views of childhood and children's vulnerability underpinning UNHCR's priorities and responses.

Implications for policy and practice

Our findings highlight the need for researchers, policymakers and practitioners to spend time with young people living in difficult situations in order to understand how they perceive the risks they face and what choices and solutions are available to them. Despite UNHCR's attempts to deploy the Age, Gender and Diversity Mainstreaming processes described above, the approach is inhibited by the limited time staff members can spend with particular refugee communities, and the challenge of incorporating this research into child protection policy and action. A more holistic and contextual approach would include consideration of how diversity in culturally valued child rearing practices, socio-economic realities, and political contexts produce varying strategies to keep children safe from harm; it would also include a range of understandings of what the best interests of a child may be.

In the Bhutanese refugee context, while UNHCR's primary response to the situation of unaccompanied and separated children is the process to determine their long-term best interests, the interventions proposed by the young researchers focused on practical and social concerns. Their recommendations included the provision of material and financial support for children without parents. Young refugees also suggested raising awareness in the camps of the difficulties faced by children living without one or both parents, and to reduce discrimination against them. This might persuade peers, friends, teachers and neighbours to offer support instead of disparagement. Young people suggested community members should be informed of the potentially negative impact of early marriage on young women's health and on both husbands' and wives' educational achievement including by adding topics on marriage, sexual and reproductive health, and family planning to the school curriculum.

A more holistic approach to protection for Bhutanese refugee children could involve increased dialogue on how to respond to difficulties experienced by young people amongst the community, and with national children's organisations (in Nepal or in other refugee hosting countries) which may improve contextual understandings of child protection. Additionally, the support of the host government, in this case the Government of Nepal, would be integral to the development of any system aiming to improve the well-being of young Bhutanese refugees living in this country.

However, the development of new systems within UNHCR to improve tracking, monitoring, and responses to children at risk appears to have occurred in the absence of adequate engagement with local communities on their understandings of childhood and child protection, but are instead guided by standardised categories of vulnerability. We suggest that UNHCR and implementing partners should invest more time in developing contextual knowledge of child protection issues and responses, which draws on the perceptions and social resources of young refugees and community members rather than conforming to pre-defined concepts of groups at risk. We suggest that further research is needed on the impact of UNHCR interventions in other settings to develop systems of child protection that respond to local contexts.

We therefore propose a review of the manner in which resources are currently allocated in child protection efforts and suggest that consideration be given instead to the above initiatives. This is in light of the considerable resources UNHCR invests in determination of best interests in certain field locations. Our research questions whether this process which relies on a particular understanding of childhood and the role of biological parents is efficient and fair. It privileges the needs of unaccompanied and separated children, which may result in the protection concerns of other young people remaining unaddressed. We suggest that some of these resources may be better spent in creating systems to support more refugee children and/or engage refugee communities in creating their own, culturally appropriate protection of children.

References

Cunningham, H. (1998) 'Histories of childhood', *American Historical Review* 103 (4): 1195–1208.

Hutt, M. (ed.) (2003) *Unbecoming citizens*, Oxford, UK: Oxford University Press.

Kinch, T. (2008) 'Linking invisibility and vulnerability: strengthening refugee child protection capacity: the case of unaccompanied and separated refugee children from Bhutan living in Nepal', unpublished MA thesis, Saint Mary's University, Canada.

Talviste, V. (2009) 'Displaced Adolescent Girls' Protection: Could Casuistry be a Methodology for Humanitarians?', Refugee Studies Centre Working Paper Series 53, Oxford, UK: Oxford Department of

International Development, available at www.rsc.ox.ac.uk/publications/working-papers/RSCworkingpaper53.pdf (retrieved 12 May 2011).

Child Research Report 1 (2007) Rosalind Evans, Fieldnotes.

Child Research Report 2 (2008) Rosalind Evans, Fieldnotes.

UNHCR (1994) 'Refugee children: guidelines on protection and care', available at www.unhcr.org/refworld/docid/3ae6b3470.html (retrieved 13 May 2011).

UNHCR (2003) 'Sexual and Gender-Based Violence against Refugees, Returnees and Internally Displaced Persons: Guidelines for Prevention and Response', UNHCR (SGVB Guidelines), Geneva, Switzerland: UNHCR, available at http://www.unhcr.org/protect/PROTECTION/3f696bcc4.pdf (retrieved 25 April 2012).

UNHCR (2005) 'UNHCR's Strategy and Activities concerning Refugee Children', available at www.unhcr.org/3bb3107ba.pdf (retrieved 14 May 2011).

UNHCR (2007) 'The Executive Committee Conclusion on Children at Risk', available at www.unhcr.org/refworld/docid/49997afa27.html (retrieved 11 May 2011).

UNHCR (2008) 'UNHCR Guidelines on Determining the Best Interests of the Child' available at www.unhcr.org/refworld/docid/48480c342.html (retrieved 13 May 2011).

Rethinking orphanhood and vulnerability in Ethiopia

Gina Crivello and Nardos Chuta

'Orphans' became a category of vulnerable children deserving special protection in the context of the global AIDS epidemic, and currently the notion of 'orphans and vulnerable children' (OVC), dominates much of the policy for protecting children across sub-Saharan Africa. Analysis of survey and qualitative data from Young Lives in Ethiopia found that parental death does not guarantee the often assumed negative impacts on children's experiences, and that inequalities between children are greater along dimensions of poverty and household location, compared to orphan status. 'OVC' obscures poverty as a main source of child vulnerability and is therefore an outdated approach.

Repenser la condition d'orphelin et la vulnérabilité en Éthiopie
Les « orphelins » sont devenus une catégorie d'enfants vulnérables qui méritent de faire l'objet d'une protection spéciale dans le contexte de l'épidémie mondiale de sida et, à l'heure actuelle, la notion d'« orphelins et enfants vulnérables » (OEV) domine une grande partie des politiques générales relatives à la protection de l'enfance dans l'ensemble de l'Afrique subsaharienne. Une analyse de données qualitatives et issues d'enquêtes de Young Lives en Éthiopie a montré que la mort des parents ne garantit pas les impacts négatifs souvent supposés sur les expériences des enfants et que les inégalités entre enfants dépendent davantage de la pauvreté et de la situation géographique des ménages que de la condition d'orphelin. L'appellation d'« OEV » fait oublier que la pauvreté est une source principale de vulnérabilité des enfants et elle constitue donc une approche dépassée

Repensando a orfandade e a vulnerabilidade na Etiópia
Os "órfãos" tornaram-se uma categoria de crianças vulneráveis que merece proteção especial no contexto da epidemia global da AIDS e, atualmente, a noção de "órfãos e crianças vulneráveis" (OVC) domina grande parte da política de proteção de crianças em toda a África Subsaariana. A análise de pesquisas e dados qualitativos de Young Lives na Etiópia constatou que a morte dos pais não resulta necessariamente nos impactos negativos frequentemente presumidos sobre a experiência das crianças e que as desigualdades entre elas são maiores ao longo das dimensões da pobreza e local de moradia da família do que em relação ao status de órfão. A abordagem dos "OVC" oculta a pobreza como principal fonte da vulnerabilidade infantil e, assim, constitui uma abordagem ultrapassada.

Una nueva concepción sobre la orfandad y la vulnerabilidad en Etiopía
A partir de la epidemia mundial de SIDA, los "huérfanos" se convirtieron en una categoría de niñez vulnerable que amerita protección especial y hoy en día el concepto de "huérfanos y

niñez vulnerable" (HNV) es uno de los aspectos que más se tienen en cuenta en las políticas diseñadas para la protección de la niñez en el África subsahariana. Un análisis de las encuestas y los datos cualitativos recopilados por Vidas Jóvenes de Etiopía demuestra que la muerte de los padres no siempre produce el impacto negativo en la vida de los niños y niñas que comúnmente se presupone. Asimismo, las desigualdades que existen entre los niños se manifiestan más por la pobreza y por la ubicación de sus hogares que por ser huérfanos. La categoría HNV oculta el hecho de que la pobreza es una de las razones más importantes de la vulnerabilidad de la niñez y por tanto ha dejado de tener vigencia.

Introduction

In recent decades, the image of the orphan has come to embody child vulnerability in sub-Saharan Africa. Estimates of the alarming numbers of children being orphaned by AIDS-related deaths in the region raised international concern about an 'orphan crisis' and the burden of their care in already fragile family and community circumstances (Abebe and Aase 2007). In 2003, the publication of *Africa's Orphaned Generation* warned,

> *with the number of children that require protection and support soaring – and ever-larger numbers of adults falling sick with HIV/AIDS – many extended family networks have simply been overwhelmed.*

Thus, '*the children of Africa, orphaned by HIV/AIDS, are the world's responsibility*' (UNICEF 2003: 6–7).

The care and protection of orphans features prominently in the agendas of key international agencies and donors seeking to improve child well-being globally; and they are behind the definitions, priorities, and solutions guiding national plans to address child vulnerability. For example, one of the leading international agencies defined an orphan as a child who has lost one or both parents, and this technical definition has been adopted widely and at many levels for policy, programming, and research worldwide (e.g., USAID, UNICEF, and UNAIDS 2002).

The attention given to orphans in the international child protection discourse suggests that orphanhood is a major, if not *the* major factor affecting child vulnerability in sub-Saharan Africa. An active research agenda focused on the risks associated with orphanhood has developed in tandem with a growing policy concern and programmatic focus targeting orphans and vulnerable children (OVC), throughout the region. However, the very concept of OVC illustrates the tension that exists between targeting specific groups of children for support – such as orphans – and developing strategies for addressing child vulnerability more generally.

This article examines some of these challenges with reference to the case of Ethiopia, where an OVC approach guides child protection. We use evidence from Young Lives,[1] a long-term study of child poverty that has been tracking two age cohorts of children, and their families, in Ethiopia since 2000. Survey analysis examined the degree to which child development outcomes were associated with orphan status, and other factors, such as household poverty. Analysis of qualitative data focused on children's own views and experiences of parental death within the context of their everyday lives. We situate our findings within the ongoing policy debates and conclude with a call to re-think the assumptions guiding OVC approaches in Ethiopia and, more broadly, the validity of imposing universal definitions on local experiences of child vulnerability.

The global spotlight on orphans

That orphans have emerged as a significant international child protection concern is illustrated by the results of a search of UNICEF's publications database covering the period 1996–2010 (UNICEF.org 2011). A search for publications tagged with 'orphan' resulted in 49 hits for the first five year period (two publications in 1996), rising to 610 results in the year 2010. During this period, the orphan problem remained rooted in concerns about the impacts of HIV/AIDS.

At the height of the AIDS crisis, '*AIDS orphans' became a new category of 'vulnerable children' requiring special protection and attention*' (Bray 2003: 52), but the term eventually fell out of favour because it was potentially stigmatising and failed to acknowledge that children could be negatively affected by AIDS regardless of orphan status. A joint report produced by UNICEF, UNAIDS, and USAID explained, "*While not all orphaning is due to HIV/AIDS, orphaning remains the most visible, extensive, and measureable impact of AIDS on children*" (USAID, UNICEF, and UNAIDS 2004: 4). New terms like 'orphans and other children made vulnerable by HIV/AIDS' reflected these concerns, broadened beyond orphans, but remained AIDS-focused.

The alternative designation of OVC leaves out the reference to AIDS, and as an international framework gained in popularity and use in the early 2000s. Orphans are conspicuously the only named category of children, and despite disentangling itself in name from the AIDS epidemic, it remains strongly rooted in the global AIDS agenda. For example, in the most recent report on *The State of the World's Children*, the subject of orphans appeared in sections on the risks of HIV, national AIDS strategies and reproductive health (UNICEF 2011). On the websites of USAID and Save the Children USA, information on OVC was accessed via sections on HIV/AIDS.[2]

The visibility of orphans in agency and donor agendas has been influential in shaping national plans of action for targeting child vulnerability throughout sub-Saharan Africa. For example, Ethiopia's first National Plan of Action on OVC (2004–06) was developed as part of a Rapid Country Assessment, Analysis, and Action Planning (RAAAP) across 17 countries.[3] However, the research base supporting such plans is flawed, lacking analytical clarity and consistency, including in basic definitions, such as who is defined and counted as an orphan (Bray 2003; Henderson 2006; Meintjes and Giese 2006; Richter and Desmond 2008; Sherr *et al.* 2008). This has led to contradictory findings regarding orphanhood as a source of child vulnerability.

The research and policy literature has also raised concerns about the ways in which orphans are conceptualised and how conceptualisations affect family and social cohesion, and the way young people see themselves (that is, through potentially exacerbating stigma or privilege) (Meintjes and Giese 2006; Sherr *et al.* 2008). In some cases, the orphan label has become a privileged identity and a way to access aid from donors (Cheney 2010; Freidus 2010). Despite confused messages and unintended consequences, there is a substantial amount of international aid available for OVC support via HIV/AIDS funding; maintaining a focus on orphans may be an effective strategy for eliciting emotive responses and for mobilising donor attention and funds (Meintjes and Geise 2006: 416).

The case of the US Government is illustrative, beginning in 2003, when PEPFAR – the US President's Emergency Plan for AIDS Relief – was launched to combat global HIV/AIDS. In 2008, nearly five billion dollars was earmarked over a five year period for OVC, and in 2011,

Ethiopia received funding for a new 100 million dollar programme to assist an estimated 500,000 orphans, vulnerable children, and families every year over the next five years.

OVC in Ethiopia

The Ethiopian Ministry of Health recorded 2,656,195 single and double orphans in the country in 1987, and the numbers have increased to reach over five million children under 15 years old (EMOH 2007). Approximately 20 per cent of orphans are believed to have lost parents as a result of HIV/AIDS (EMOH 2007; UNAIDS 2008). HIV/AIDS is a leading cause of orphanhood in Ethiopia, although the disease does not presently account for the majority of parents' deaths. In addition, there have been important episodic causes of orphanhood or separation of children from their parents, including famine, conflict, and displacement (Cooper 2010: 10). Orphanhood and parental absence are structural features of Ethiopian society (Abebe and Aase 2007: 2059).

High rates of adult mortality in Ethiopia coincide with widespread poverty. Ethiopia is one of the world's poorest nations and in 2011 was ranked 174 out of 187 countries on the Human Development Index (UNDP 2011). Ethiopia's Household Income and Consumption Expenditure survey (2004/5) identified 38.7 per cent of the country's population as poor, with higher rates of poverty in rural areas where approximately 84 per cent of the population lives (Woldehanna *et al.* 2011: 17).

As in other countries across the region, Ethiopia adopted an OVC framework to guide action for vulnerable children, and there are currently plans to carry out a new situation analysis of OVC across the country in order to inform a new National Plan of Action for OVC. A new National Plan of Action for Children is being developed simultaneously, raising concerns amongst critics who question the need for separate plans and who point out that policy makers do not use existing research on OVC, so generating more data is unnecessary.

PEPFAR funded development of the country's 'Standard Service Delivery Guidelines for Orphans and Vulnerable Children's Care and Support Programs' (2010) which included definitions for OVC, defining an orphan as:

> *a child who is less than 18 years old and who has lost one or both parents, regardless of the cause of the loss. A vulnerable child is a child who is less than 18 years of age and whose survival, care, protection or development might have been jeopardized due to a particular condition, and who is found in a situation that precludes the fulfilment of his or her rights.* (MOWA and FHAPCO 2010: 3)

While the guidelines go some way towards defining child vulnerability within the country, such definitions reflect international conceptualisations of childhood as defined chronologically below the age of 18, and global notions of child vulnerability based on threats to individual children's rights and interpersonal abuses. Stronger links are therefore necessary between research, policy, and practice on the one hand, and, on the other, the contextual factors often lacking in such bureaucratic definitions. The question for this paper is: does the focus on orphanhood reflect the realities of Ethiopian children?

Context of this research

We analysed data from Young Lives in Ethiopia: the third round of surveys (2009) and a qualitative sub-study (2010) focusing on children's experiences of orphanhood and vulnerability. Sample households were poorer than average Ethiopian households but overall represented a wide range of poverty circumstances, albeit not nationally representative.

Orphanhood was not a criterion for recruiting study participants; nonetheless, we found that by the time they were aged 15, around 21 per cent of the older children had lost one or both parents, and among the younger group, aged eight, the figure was 10 per cent. Nationally, an estimated 18 per cent of Ethiopian households include at least one child who has lost one or both parents. At the time of the last survey (2009), the younger children in our sample were aged seven to eight (n = 1882) and the older children were aged 14–15 (n = 973). 5.8 per cent of the younger children and 14 per cent of the older children were paternal orphans; reported maternal orphanhood was two and four per cent, respectively. Parental absence[4] was even more prevalent than parental death: 19 per cent of the younger cohort and 22 per cent of the older cohort were reported as having missing fathers. Thus, paternal death was the most common form of parental death, and when a parent was reported absent, it was also more likely to be the father.

Participants in the qualitative sub-study were drawn from the existing Young Lives sample located in Addis Ababa and Hawassa cities. Data collection involved 63 individual interviews with 26 focus-children aged 9–16, caregivers, NGO and local government representatives, teachers, healthcare providers, and *Iddir* (traditional burial association) leaders, and separate group discussions with children and caregivers. Prior to fieldwork, a policy context analysis involved 17 semi-structured interviews with representatives from relevant NGOs, government, and international organisations working in the area of OVC in Addis Ababa.

Qualitative fieldwork took place in Bertukan (Addis Ababa) and Leku (Hawassa). Bertukan is a hub for commerce and industrial activities, with a population of over 14,000. It is regarded as the city's old quarter, and poverty is widespread. Street vending of fruits and vegetables, renting houses for storage, and carrying goods for cash are key sources of income for residents. There are many problems in the locale related to housing conditions, sanitation and hygiene, infrastructure, economy, and societal risks.

Leku is an active quarter of Hawassa, the capital of the Southern Nations, Nationalities, and Peoples Region (SNNPR), with roughly 23,000 inhabitants. Poverty is widespread and residents earn their living through petty trade, street vending, and daily labour. Common risks faced by children in the area included the presence of *shisha* houses (places where people go to smoke and chew *chat*, a mild stimulant leaf), bars, and brothels, and overcrowded living conditions.

Survey results

Our findings raise several questions about the relationship between orphanhood and vulnerability within the context of widespread poverty and rural/urban inequalities, both in terms of children's outcomes and their experiences. We begin by looking at the care of orphans in our sample, which is a major policy and programmatic concern in Ethiopia. Children who had experienced the death of one or both parents were more likely to be living in better-off households, this possibly reflecting a tendency for children to be taken in by family members who were financially better-off and the greater prevalence of parental death in urban settings. We found that despite considerable parental death and absence in our sample households, most children were reported to be living with relatives: only two per cent of the older cohort and 0.4 per cent of the younger cohort were living with non-relatives.

Living arrangements differed depending on whether it was the mother or the father who had died or had been reported absent. There was a greater diversity of caregivers for children whose mothers had died or been reported absent. If the father had died or was absent, the caregiver was likely to be the biological mother (89 per cent younger cohort; 90 per cent older cohort). In mothers' absence, grandmothers tended to take on caring responsibilities. Care arrangements were also shaped by the age of the children, such that younger maternal orphans were more likely to be living with their fathers (43 per cent) when compared to older children whose mothers had died (22 per cent).

For children who had lost both parents, patterns of care also differed by age cohort; younger children in this group were more likely to be cared for by a grandparent (64 per cent, compared to 20 per cent of older 'double orphans'); a much greater diversity of caregivers were reported for older double orphans. Analysis of Ethiopia's Demographic Health Surveys of 2000 and 2005 reported similar trends, finding that 83 per cent of paternal orphans lived with their mothers, while 70 per cent of maternal orphans lived with their fathers, and 51 per cent of double orphans lived with a grandparent (Beegle *et al.* 2010).

Children's development outcomes

A widespread policy and programming assumption has been that development outcomes for orphans are different relative to other children, and that orphans fare worse. We explored this assumption on a selection of survey variables relating to schooling and health outcomes. We found that younger orphans (single or double) were slightly *more* likely to be enrolled in school when compared to younger children whose parents were alive. This was in contrast to older orphans who were slightly *less* likely to be enrolled in school than other children their age. For both age groups, double orphans were more likely than any other group to be enrolled in school (96 per cent older double orphans, compared to 90.6 per cent non-orphans; 92.9 per cent younger double orphans, compared to 76.5 per cent non-orphans). We also found that for both age groups orphan status did not account for significant differences in reports of missing (a week of) school in the previous year.

There were, however, disparities in school enrolment amongst both age groups, depending on household location, favouring urban children. Similar disparities were evident in terms of household wealth, with children in better-off households more likely to be enrolled in school, for both age groups. The survey also found that in households reporting long-term illness children were more likely to have missed school (younger cohort 17.7 per cent; older cohort 14.8 per cent) compared to children in households without long-term illness (younger cohort 9.8 per cent; older cohort 11.6 per cent).

Information on body mass index (BMI) as an indicator of health for the older children found that overall 36 per cent of the cohort was below the healthy range. There were minor differences in BMI (older cohort) and height-for-age (younger cohort) in relation to orphan status. There were greater disparities based on where children lived, with children in rural households less likely to hit the healthy range. The wealth status of households was also associated with health outcomes such that children in poorer households were less likely to fall within the healthy range according to these measures.

The sample was then divided by orphan status, focusing on those households in the two poorest and two richest categories. Orphans living in better-off households fared better on health indicators compared to orphans in the poorest households. Results for the older cohort indicated that orphans in the poorest quintiles were doing better than non-orphans in the poorest quintiles (65 per cent and 55 per cent within the healthy range for BMI, respectively). In the younger cohort, there were no differences between orphans and non-orphans in the poorest households.

The survey findings indicated that inequalities in schooling and health outcomes were larger along other dimensions of vulnerability, such as the location of children's households (rural or urban) and the material circumstances of their households. These trends are in line with findings from other studies comparing development outcomes for orphaned children relative to other children (Bray 2003; Campbell *et al.* 2008; Richter and Desmond 2008). Children's outcomes, however, are not the same as their subjective experiences, and it is important to understand the ways in which young people experience and make sense of parental loss within the context of their everyday lives.

Children's experiences

Daily life in Bertukan and Leku is fraught with difficulties for both young and old. We analysed children's experiences along three main axes: the first pertained to children's care relationships, the second to their social and material contexts, and the third to the ways in which young people made sense of parental death.

Over half of the children involved in the qualitative study had experienced some form of parental death, and only two of the 25 children reported having both parents alive and living with them. In other cases, one or both parents were alive but living apart from the child, or perhaps nearby but not with the child due to housing shortage; some had migrated for work and a few had abandoned their children.

Children's care arrangements were therefore diverse, the majority of children lived with relatives, and even where parental death and absence were reported, many remained living with their mothers (father's death or absence being more prevalent). Regardless of who was the primary caregiver, the quality of children's care and the security of their relationships were important factors shaping their experiences of orphanhood and vulnerability.

Children's roles within the burden of their care

Given the fragility of many Ethiopian households, the role of the extended family and of community-based organisations has become a key question and OVC programmatic focus. Historically, in many African societies, the extended family played an important function in raising the young – including when parents were living – through short or long-term fostering, as a response to household crisis, as well as under favourable conditions (Abebe and Aase 2007; Abebe and Skovdal 2010; Kayongo-Male 1984). Amongst our sample households, family and other informal networks were crucial for support and survival. Household membership was fluid and many children resided with extended family. This was due to poverty and poor housing, parental migration for work, parental death or abandonment, and better opportunities for children (such as higher quality schooling).

A main policy concern has been the sustainability of this traditional form of protection and whether the numbers of orphans requiring homes has pushed families beyond their capacity (Abebe and Aase 2007). Such is the burden of orphan care and the focus on protecting vulnerable children. Whilst acknowledging that all vulnerable children should be supported, it is important also to affirm the valuable ways in which young people contribute to the care networks of which they are a part. An examination of the experiences of orphanhood within the 'mutuality' of children's care relationships (Abebe and Skovdal 2010) shifts the focus away from what children lack (namely parents) and towards the resources they bring to bear on their situations.

For example, 15-year-old Denbel was living with his mother and brother in Hawassa in extreme poverty. He had no place at home to study, and during the rainy season, the corrugated iron roof leaked and soiled his school notebooks. His mother earned money by trading firewood and Denbel combined daily wage labour with work in a local shop, managing both jobs alongside schooling. Denbel's father died when he was two years old, and the family's economic situation deteriorated.

> *My mother always says that she doesn't have the capacity to fulfil all our needs. And I don't feel good because I don't have a father.*

He got a job, explaining, '*when my family couldn't provide me with the necessities, I decided to work on my own and earn money to get what I need*'. He gives his mother about a third of his earnings and uses the rest to cover his personal expenses.

Denbel believed that his family would not be struggling so, had his father been alive. But he accepted death as a natural and irrevocable occurrence, and did not think that he would suffer long-term effects.

> *You know, we all are created to die. So, I don't have to remember him all my life since I myself will die one day.*

Denbel's case fits the sense of respect and obligation that structures children's roles and their responsibilities towards their households (Boyden 2009; Poluha 2004). Another example, of 15-year-old Bereket, showed how children's work also contributed to a sense of pride and hope for the future. Bereket was being raised by his 60-year-old grandmother in Addis Ababa – along with three brothers and two cousins. After the school day ended, Bereket worked washing cars and changing tires, and he was proud of what he learned at work:

> *I can manoeuvre some garage machines. I have a little driving skill, in addition to tire repair and maintenance ... I take risks to learn new things ... I still have some friends who don't know how to drive because they don't take risks.*

Although Bereket and Denbel faced multiple layers of disadvantage, neither saw himself as a dependent victim. Their capacity to work and to earn money nurtured their sense of hope and contributed to their daily survival. Boys' and girls' experiences of orphanhood were therefore grounded within the context of their everyday roles, responsibilities, and relationships. Being male was an important source of advantage for Denbel and Bereket in this respect, opening up opportunities for their apprenticeships and for mastering the skills and knowledge facilitating their adult roles later in life. Petty jobs like working in the fruit and vegetable markets, grocery stores, or mechanic shops were often more accessible to boys, while girls' work clustered around or inside the home. Nonetheless, girls' contributions were also essential for mitigating household poverty and the effects of parental death, as part of mutual systems of care.

Orphanhood and everyday risk

A holistic view of orphanhood considers the wider social and material contexts of children's daily lives, including the pressing, everyday risks that shape childhood in these contexts. Crucial among these were the ways in which gender, age, and household poverty interacted in shaping children's vulnerabilities. Though gender can be an advantage, it can also be a risk. Gender-based violence, such as rape and harassment, were common fears for teenage girls, and this influenced their livelihood opportunities and feelings of insecurity in the streets. Crowded living conditions were also common and brought together strangers renting accommodation under the same roof. This presented risks for girls and their moral standing, such as inappropriate or early sexual relations, often initiated through the exchange of favours and material goods.

These were clear concerns for 15-year-old Fatuma's mother, in Addis Ababa. Fatuma's father died when she was very young and she had little memory of him. Fatuma worked a lot at home. She made *injera* (a traditional bread made of teff flour) to sell, especially when the mother became ill. Caregiver illness was common among households in our study, and frequently had economic repercussions related to diminished earning capacity and costs for treatment, on top of psychosocial impacts and anxieties about future uncertainty.

Fatuma's mother's life had been fraught with difficulties, even before becoming ill. She never attended school and when she went to work as a cleaner in a widow's home she was raped by one of the boarders. She became pregnant with a son who was subsequently raised by an aunt in the village. When she was pregnant with Fatuma, her husband, a different man, abandoned her, only to return five years later. Her mother remembered when Fatuma was born: '*I had nothing to cover her body with, so I used long sleeves of my clothes and made a trouser and a dress out of it*'. Even now, she struggled to make ends meet as the money from washing clothes was not enough for the family.

Renting out space in their house was therefore crucial, but a significant source of worry for the mother. One boarder, a young day labourer, concerned her:

> *The boy has a mobile phone and wants to lend it to Fatuma, but sometimes, both of them quarrel and I don't know the reason and she does not tell me either ... I am not comfortable with the way the boy is approaching my daughter.*

When asked about what she considered the main problems in her life, Fatuma made no mention of the young man, and instead pointed to her mother's illness and her fears of losing her. Nor did she emphasise poverty as a major obstacle:

> *I don't think that poverty will be with me forever. So I am not worried about it ... If I work hard, there is no reason why I should not fulfil my future goals the same as my friends and even reach a better position.*

Tsega's case also illustrates this theme. She was 15 years old and living with her mother and eight siblings in Hawassa. Tsega's father died when she was 11, and following his death, her mother and older brother sought paid work, but her mother's job as a dishwasher and cook was frequently disrupted due to chronic illness.

The family's recent problems began when the father was crushed by a car while riding a bicycle, followed by two years of intense medical treatment; one of his legs became infected and he died. Prior to this, the family lived relatively comfortably; they owned their house, several horse drawn carts, and a hotel and butchery house. One by one these were sold to finance the father's treatment. Financial difficulties resulted in Tsega enrolling late in school and repeating grades. Echoing what many other children her age have said, and in spite of the difficulties facing the family, Tsega stated, '*as man is mortal, he [her father] died and so I have to accept*'.

One of Tsega's biggest worries is being raped in her neighbourhood. Her sister had been cheated by a boy promising marriage, but who abandoned her when she fell pregnant. Now that Tsega was a teenager, her mother feared that Tsega might also be cheated by a young man promising her things that she herself could not afford to provide for her daughter.

Indeed, children's vulnerabilities were shaped by their household and community circumstances, rooted in poverty, and exacerbated by parental illness and death, migration, family separation, and gender dynamics. Orphanhood within this context became one of many sources of vulnerability and not often the most pressing daily concern for boys and girls. Daily life was further shaped by intergenerational processes whereby long histories of parental disadvantage played out in children's current realities, providing, on the one hand, valuable lessons for the young, and on the other hand, another layer of hardship and risk to overcome.

Making sense of orphanhood

Many young people lacked detailed knowledge of their parents' passing, especially if they were young when their parents had died. It was also common for adults to shield children from painful information, so children's memories were often vague. These factors influenced their explanations and descriptions of parental loss, as did the quality of their current family relationships and household circumstances. In other words, where care and support were strong, children did not refer to parental death as a main factor shaping their current lives.

The case of 15-year-old Genet in Addis Ababa is illustrative. She began living with her aunt following the death of her mother (when she was five) and two years later her father, who had endured long periods of illness. In Genet's view, her aunt and uncle provided well for her, and theirs was amongst the better-off households in our sample. When asked about missing her parents she replied, '*I feel nothing because I don't know them very well ... I feel a little sadness but not so much because I have missed nothing*'. The material security and sense of nurture experienced by Genet mediated her experiences of parental death.

This was not the case for 14-year-old Tagesu in Hawassa, who explained much of her current difficulties with reference to her father's death. Her family history was full of gender-based violence. Tagesu's mother had married by abduction[5] at the age of 15, and after giving birth to two children, she divorced. She later married Tagesu's father, but he died when Tagesu was three years old. When the mother migrated in search of work leaving behind Tagesu and her elder sister to work as day labourers, the sister was raped by a man whom she later married and whose child she bore.

Tagesu believed that her father's death plunged the family into deep poverty and deprived her of his love and protection. Many of her needs remained unmet, and she felt that had her father been alive he would have educated her in a private school and given her plenty of food.

> *It is the hunger and lack of food that I cannot handle. I tell my mother when I am hungry. If she has something, she gives it to me, and if she does not have anything she weeps with me.*

In contrast to Genet's account, Tagesu and her family contended with multiple layers of disadvantage and the nature of risk for their household was cumulative and compounding, influencing the meaning Tagesu gave to her father's death.

Many of the older children expressed sadness regarding parental loss, but their views also demonstrated a degree of fatalism which may have aided their coping over time. They considered parental death inevitable and natural. For many of them, like Tagesu, daily life entailed confronting a constellation of risks within and outside of the household. Poverty was fundamental in shaping the way they experienced a variety of shocks, including parental death, absence, abandonment, and illness. Their struggles and successes were lived out within the context of mutual care systems wherein children's contributions were often essential for household survival. Under these circumstances, the extended family and community comprised a major resource for children, particularly in the absence of external support, and their positive relationships with caregivers were protective in mediating their experiences. In the absence of a strong web of care and material support, parental death provided children with the explanation that gave meaning to their daily struggles.

Conclusion

The notion of OVC represents a positive move towards recognition that orphaning is not the sole source of risk for children in sub-Saharan Africa, where the label has most consistently been applied. Orphans within the context of the AIDS epidemic have, nonetheless, retained a prominent place in OVC discourse at many levels – from international agency and donor agendas, through national-level OVC action plans, down to programmes – and in some contexts, the OVC identity has become a new social category and personal identity. Thus, OVC is not simply an innocuous bureaucratic label created to measure parental death and child vulnerability across diverse contexts; it also shapes thinking about who the world's vulnerable children are, and to funnel global funds to support them.

In Ethiopia, the term OVC is widely used by the government and other stakeholders, including development practitioners, and to date, the category has not been widely questioned. However, increasing research evidence indicates a frequent mismatch between externally-generated bureaucratic categories such as 'orphan' and 'OVC', and locally-derived understandings of parental loss, family, childhood, or vulnerability. As happened to the categories that came before it, cracks in the OVC framework are beginning to show at international level.[6]

Researchers and policymakers in Ethiopia must therefore decide whether their efforts will become more targeted or more universalised. That the Ethiopian government is simultaneously developing separate plans of action for OVC and for the country's general child population epitomises this dilemma. In line with other studies, our research findings from Ethiopia suggest that a focus on orphans obscures the vulnerabilities of other groups of children. We presented survey analysis indicating that inequalities in schooling and health outcomes were larger along other dimensions of vulnerability, such as the location of children's households and the material circumstances of their households, when compared to analysis by orphan status. Vulnerability was related to living in the context of acute poverty, so childhood poverty and associated risks need to be tackled, rather than specifically targeting orphans as a policy/programmatic priority.

Qualitative analysis also showed that orphanhood does not guarantee negative impacts on children's experiences and development outcomes, and many of the problems faced by orphans are also faced by other disadvantaged children in their communities. Indeed, many of the types of aid provided to orphaned children target their poverty (food, school materials, and clothing); it is therefore important to address universal access to basic supports for children, such as through social protection programmes.

Rethinking orphanhood and vulnerability also requires improved understanding of risk factors for children, and further research is needed on the interaction and accumulation of risks for children, on the mediating factors that decrease or enhance child vulnerability, and on the importance of contextual factors. In this study, several factors were shown to influence young people's experiences of parental death, and the quality of children's care relationships and their material security were crucial among these. However, poverty and gender-based violence were among the more pressing everyday risks for many children. The meanings and experiences of childhood, family, vulnerability, and orphanhood are highly contextual, so greater attention to the social dimensions of child vulnerability is needed. In this respect, policy for protecting children in Ethiopia must also acknowledge the resources that young people bring to bear on their household circumstances through their agency, roles, and responsibilities. The category of OVC diverts attention away from the crushing poverty that is the backdrop for the childhood of many in Ethiopia. Tackling childhood poverty needs to be at the heart of strategies to address the needs of *all* vulnerable children currently growing up in the country.

Acknowledgements

We are grateful to the children, families and others who collaborate with Young Lives. Yisak Tafere led the fieldwork and reporting for this study. Elizabeth Cooper, Marlijn Lelieveld, Claudia Seymour, and Laura Valadez provided excellent research consultancy. Helen Murray reviewed an earlier draft of the paper. Young Lives is core-funded from 2001 to 2017 by the UK Department for International Development (DFID), and co-funded by the Netherlands Ministry of Foreign Affairs from 2010 to 2014. Bernard van Leer Foundation and the Oak Foundation fund sub-studies, and the latter funded the study on orphanhood and vulnerability in Ethiopia.

Notes

1. See www.younglives.org.uk for research design and methodology.
2. Accessed 20 July 2011 and 7 September 2011, respectively.
3. The exercise was created by USAID, UNICEF, UNAIDS, and The World Food Programme.
4. The survey asks how often the child sees his/her biological parents. 'Absence' here captures anything from 'sees weekly' to total absence.
5. 'Abduction' is a type of marriage practiced in rural areas when the boy is poor (thus avoiding payment of bridewealth) or when the girl refuses to accept a marriage proposal.
6. See the 2010 report to the US Congress on the Public Law 109-95 (The Assistance for Orphans and Other Vulnerable Children in Developing Countries Act of 2005) which shifted focus away from OVC and towards 'highly vulnerable children'.

References

Abebe, T. and A. Aase (2007) 'Children, AIDS and the politics of orphan care in Ethiopia: the extended family revisited', *Social Science and Medicine* 64 (10): 2058–69.

Abebe, T. and M. Skovdal (2010) 'Livelihoods, care and the familial relations of orphans in eastern Africa', *AIDS Care* 22 (5): 570–6.

Beegle, K., D. Filmer, A. Stokes, and L. Tiererova (2010) 'Orphanhood and the living arrangements of children in sub-Saharan Africa', *World Development* 38 (12): 1727–46.

Boyden, J. (2009) 'Risk and capability in the context of adversity: children's contributions to household livelihoods in Ethiopia', *Children, Youth and Environments* 19 (2): 111–37.

Bray, R. (2003) 'Predicting the social consequences of orphanhood in South Africa', *African Journal of AIDS Research* 2 (1): 39–55.

Campbell, P., S. Handa, M. Moroni, S. Odongo, and T. Palermo (2008) 'Assessing the "Orphan Effect" in Determining Development Outcomes for Children in 11 Eastern and Southern African countries', Nairobi, Kenya: UNICEF.

Cheney, K. (2010) 'Expanding vulnerability, dwindling resources: implications for orphaned futures in Uganda', *Childhood in Africa* 2 (1): 8–15.

Cooper, E. (2010) Orphans and Vulnerable Children and Risk in Ethiopia: A Review of Research Findings', unpublished report', Oxford, UK: Young Lives.

EMOH (2007) *AIDS in Ethiopia, 6th Report*, Addis Ababa, Ethiopia: Ethiopian Federal Ministry of Health/National HIV and AIDS Prevention Control Office, available at www.etharc.org/AIDSinEth/publications/AIDSinEth6th_en.pdf (retrieved 27 February 2012).

Freidus, A. (2010) 'Raising Malawi's children: unanticipated outcomes associated with institutionalised care', *Children & Society* 24 (4): 293–303.

Henderson, P. C. (2006) 'South African AIDS orphans: Examining assumptions around vulnerability from the perspective of rural children and youth', *Childhood* 13 (3): 303–27.

Kayongo-Male, D. (1984) *The Sociology of the African Family*, New York, USA: Longman.

Meintjes, H. and S. Giese (2006) 'Spinning the epidemic: the making of mythologies of orphanhood in the context of AIDS', *Childhood* 13 (3): 407–30.

MOWA and FHAPCO (2010) 'Standard Service Delivery Guidelines for orphans and vulnerable children's care and support programs', Addis Ababa, Ethiopia: Ministry of Women's Affairs and Federal HIV/AIDS Prevention and Control Office, available at http://hivaidsclearinghouse.unesco.org/search/resources/iiep_ethiopia_ovc_guidelines_2010.pdf (retrieved 27 February 2012).

Poluha, E. (2004) *The Power of Continuity: Ethiopia through the Eyes of its Children*, Uppsala, Sweden: Nordiska Afrikainstitutet.

Richter, L. and C. Desmond (2008) 'Targeting AIDS orphans and child-headed households? A perspective from national surveys in South Africa, 1995–2005', *AIDS Care* 20 (9): 1019–28.

Sherr, L., R. Varrall, J. Mueller, L. Richter, A. Wakhweya, M. Adato, M. Belsey, U. Chandan, S. Drimie, V. Haour-Knipe, M. Hosegood, J. Kimou, S. Madhavan, V. Mathambo, and C. Desmond (2008) 'A systematic review on the meaning of the concept "AIDS orphan": confusion over definitions and implications for care', *AIDS Care* 20 (5): 527–36.

UNAIDS (2008) 'Epidemiological Fact Sheet on HIV and AIDS, 2008', available at www.unaids.org/en/Regionscountries/Countries/Ethiopia/ (retrieved 17 July 2011).

UNAIDS, UNICEF, and USAID (2004) 'Children on the Brink 2004: A Joint Report on New Orphan Estimates and a Framework for Action', Washington DC, USA: UNAIDS, UNICEF, and USAID, available at http://data.unaids.org/publications/External-Documents/unicef_childrenonthebrink2004_en.pdf (retrieved 27 February 2012).

UNDP (2011) 'International Human Development Indicators: Ethiopia', available at hdrstats.undp.org/en/countries/profiles/ETH.html (retrieved 27 February 2012).

UNICEF (2003) 'Africa's orphaned generation', New York, USA: UNICEF, available a www.unicef.org/sowc06/pdfs/africas_orphans.pdf (retrieved 27 February 2012).

UNICEF (2011) 'The state of the world's children', New York, USA: UNICEF, available at http://www.unicef.org/sowc2011/fullreport.php (retrieved 27 February 2012).

UNICEF.org (2011) 'UNICEF website', available at www.unicef.org (retrieved 17 July 2011).

USAID, UNICEF, and UNAIDS (2002) 'Children on the Brink 2002: A Joint Report on Orphan Estimates and Program Strategies', Washington DC, USA: USAID, UNICEF, and UNAIDS, available at http://data.unaids.org/Topics/Young-People/childrenonthebrink_en.pdf (retrieved 27 February 2012).

Woldehanna, T., R. Gudisa, Y. Denbel, and A. Pankhurst (2011) 'Young Lives: Round 3 Survey Report Ethiopia', Oxford, UK: Young Lives, available at www.younglives.org.uk/files/presentations/woldehanna_ethiopiaround3survey-overview_27sept2011 (retrieved 27 February 2012).

Children's responses to risk in agricultural work in Andhra Pradesh, India

Virginia Morrow and Uma Vennam

This paper discusses protection of children from hazards in agricultural work. International and national policies aim to protect children by eliminating all child labour. Previous literature on hazardous child labour tends to focus on single industries or crops, overlooking the variety of activities that children undertake in subsistence farming. We analyse survey and qualitative data from children, and present rates of work, injuries experienced, how children deal with risks, and perceived benefits of work. The most effective form of prevention may be to build on existing knowledge and experience, working with communities to develop strategies to make work safer.

Ripostes des enfants face aux risques du travail agricole en Andhra Pradesh, en Inde
Cet article traite de la protection des enfants face aux aléas du travail agricole. Les politiques internationales et nationales visent à protéger les enfants en éliminant tout le travail des enfants. Les documents publiés précédemment sur le travail dangereux des enfants tendent à se concentrer sur les industries ou les cultures, sans prêter attention à la variété d'activités que les enfants entreprennent dans l'agriculture de subsistance. Nous analysons des données qualitatives et issues d'enquêtes fournies par des enfants, ainsi que les taux actuels de travail, les blessures subies, la manière dont les enfants gèrent les risques et les avantages perçus du travail. La forme la plus efficace de prévention pourrait consister à renforcer les connaissances et les expériences existantes, en travaillant avec les communautés pour élaborer des stratégies afin d'améliorer la sécurité au travail.

As respostas das crianças ao risco no trabalho agrícola em Andhra Pradesh, Índia
Este artigo discute a questão da proteção de crianças dos perigos no trabalho agrícola. Políticas internacionais e nacionais visam proteger as crianças eliminando todo trabalho infantil. A literatura anterior sobre trabalho infantil perigoso tende a focalizar indústrias ou lavouras específicas, desconsiderando a variedade de atividades que as crianças realizam na produção agrícola de subsistência. Analisamos pesquisas e dados qualitativos de crianças e apresentamos informações sobre taxas de trabalho, ferimentos ocorridos, sobre como as crianças lidam com os riscos e os benefícios percebidos do trabalho. A forma mais efetiva de prevenção pode ser a de basear-se em conhecimento e experiência acumulados, trabalhando com comunidades para desenvolver estratégias para tornar o trabalho mais seguro.

Las reacciones de la niñez ante el riesgo del trabajo agrícola en Andhra Pradesh, India
Este ensayo analiza la protección de la niñez ante los peligros del trabajo agrícola. Las políticas a nivel nacional e internacional tienen como objetivo proteger a la niñez mediante la eliminación del trabajo infantil. Anteriormente, los estudios sobre el trabajo infantil peligroso se enfocaban en las industrias o los cultivos específicos, sin tomar en cuenta las múltiples actividades que los niños y niñas realizan en la agricultura de subsistencia. El ensayo analiza la información cualitativa y las encuestas realizadas a niños y niñas, examinando las actuales tasas de empleo, las lesiones más comunes, cómo la niñez reacciona ante el riesgo y los beneficios que ésta atribuye al trabajo. Para prevenir de una forma más eficaz, el método a utilizar tendría que construirse a partir de la experiencia y los conocimientos ya existentes, mediante el trabajo con las comunidades para crear estrategias que aseguren resultados.

Introduction

This paper contributes research on hazardous child labour in agriculture, by foregrounding children's accounts. Agriculture is the largest sector of the economy in which children work (Fassa *et al.* 2010). According to the International Labour Organisation (ILO), an estimated 60 per cent of all child workers engage in agriculture (ILO 2010). Children who live on farms undertake numerous tasks related to agricultural production, in various forms of economic relationship, from family-based subsistence farming to commercial plantations, in developed and in developing countries. Some of these tasks inevitably involve physical risks and potential harm. Yet agricultural work within families tends to fall outside national and international definitions of child labour, and in many countries is unregulated.

Much research on child labour, and consequent policy initiatives, are based on the assumption that children are passive victims requiring protection from labour, that poverty causes child labour, and that children are compelled to work by their parents. In the literature on hazardous child labour, there has been a tendency to focus on single industries or specific crops like cotton seed or cocoa, which misses the variety of activities that children undertake, particularly in subsistence and market farming. The existing literature rarely acknowledges that children learn through experience not only to work, but may also learn to manage harm associated with work. Further, children persistently work and sometimes want to work (Bourdillon *et al.* 2010). Why is this? Are campaigns against child labour overlooking the fundamental place of work in children's lives? What is lost when children are removed from work? Very little research has explored what children themselves say about hazards they encounter at work, nor how they try to perhaps manage these risks (Abebe and Bessell 2011).

In response to these questions, we draw on survey data from Young Lives in Andhra Pradesh (see preface to this volume). The paper also draws qualitative data from 42 children, collected for a sub-study.

This paper does not discuss whether children should be allowed to work or be pushed to work. Nor do we attempt to determine whether agricultural child labour should be classified as 'hazardous' work, and we do not draw on objective measurements of possible effects of work on health, which are problematic (see Fassa *et al.* 2010; Levison and Murray-Close 2005). We demonstrate how children perceive their work, and how they attempt to protect themselves at work. We suggest that more could be done to work with communities, while agriculture remains at a low level of technological development and demands hard manual labour. The paper raises questions about child protection strategies that focus solely on children's work in situations of rural poverty, such as those proposed by the ILO (ILO 2011), where families face a number of threats to general health, and suggests that greater attention be paid to how communities already attempt to protect themselves.

The next section of the paper describes recent efforts to protect children from hazardous work at international and national policy level in India. The third section describes a small-scale study that explored children's experiences of work in two communities in Andhra Pradesh, India. The final section explores the fits and misfits between international aspirations and the micro-level of children's lived realities.

The background of international and national policy

What are the key current international and national policies that attempt to protect children by controlling hazardous child labour? In May 2010, the ILO held a Global Conference entitled 'Towards a world without child labour'. The meeting agreed a 'Road Map' towards the elimination of the worst forms of child labour by 2016, in line with the two main ILO Conventions relating to child labour, Convention 138, on the minimum age for work, and Convention 182 (and related Recommendation 190). The latter prohibits '*worst forms of child labour*', which include '*work which ... is likely to harm the health, safety or morals of children*' (Art 3, d). States define 'hazardous' work, but this is problematic, since 'hazardous work' lists are often borrowed from other countries, may be out-of-date, or tackle obviously hazardous work, such as mining, but not hazardous tasks found in agriculture (ILO 2011: 42).

India has taken up this global campaign, yet has not set a general minimum age for all work, and sets a low age (14 years) for work categorised as hazardous. India has not ratified ILO conventions 138 and 182 (Ramanathan 2000). The Child Labour (Prohibition and Regulation) Act 1986 prohibits the employment of children in certain occupations and processes, but does not address family-based labour. A range of national policy interventions and proposals are currently relevant, including the Right of Children to Free and Compulsory Education Act (2009) (RTE), intended to guarantee access to primary education from the ages of five to 14 years. There are campaigns by NGOs (for example, Save the Children India, CRY) to include agricultural labour in the list of 'hazardous work'. There are initiatives to eliminate child labour, including recourse to the law to 'rescue and rehabilitate' bonded child labourers. However, prosecutions of employers are rare, and none of the legislation relates to children working with their families in agriculture. International and state level policies can seem remote from the everyday lives of children in rural areas.

Large numbers of children work in India, especially in agriculture (Subbaraman and von Witzke 2007). Children's roles and activities are structured by caste, ethnicity, and religion (see Gupta 2005). Historically, children took up the roles of their parents, working alongside them from an early age. Shirol (2000) in an ethnography in rural Karnataka conducted in the late 1980s, described the relationship between social organisation and children's work linked to the caste system. Higher caste children typically worked to inherit traditional occupations rather than for wages. Lower caste children worked mainly for wages with less emphasis on developing traditional occupations. The Government of India has introduced numerous policy initiatives to try to reduce discrimination and the effects of caste and ethnicity on children's life chances.

Evidence shows that current policies are not effective and large numbers of children continue to work (ILO 2011). We suggest that incorporating children's views about their work could enrich understanding and practice in this area. To do so would be in line with Recommendation 190 (Section 1, Para 2) of ILO Convention 182, which suggests that programmes of action should be designed '*in consultation with relevant government institutions and employers' and workers' organizations*, taking into consideration the views of the children directly affected *by the worst forms of child labour, their families and, as appropriate, other concerned groups* ...' (emphasis added).

Children's experiences of work

The next section briefly presents some findings from the Young Lives survey, before presenting findings from qualitative research relating to risks/harms at work.

How many children are working?

Young Lives collects data from 2,000 children born in 2000–01 (the younger cohort) and 1,000 children in each country born in 1994–95 (the older cohort). There have been three rounds of survey data collection in Andhra Pradesh, with the third conducted in 2009. Information on child work comes from two components of the survey, a set of questions asking children about the amount of time spent on various tasks in a 'typical day' in the week preceding the survey, and a question that asks whether children have done any paid work in the previous 12 months. There is a fairly strong correspondence between answers to these. Analysis uses paid work on a typical day, paid or unpaid work for the household (such as farm labour, but excluding domestic chores), and doing any paid work in the previous 12 months. The percentage of older cohort children who were in paid and unpaid work rose from six percent in Round 1 (2002, when the children were aged seven–eight years) to 22 percent in Round 2 (2006–07, when they were 12) to 28 percent in 2009 (when they were 14–15). In 2009, 33 percent of children reported work in rural areas compared to 12 percent in urban areas, reflecting the dominance of agricultural activities in rural children's lives. However, the time-use question may underestimate the extent of children's work, because in rural households the distinction between domestic labour and unpaid work is hazy: for some children, cattle grazing may be seen as a household chore.

By the time children reach secondary school age, many of them are expected to engage in work at home and on farms, especially during the peak agricultural seasons. Some children manage school, home and farm work in complex ways (Morrow and Vennam 2010). Children's work needs to be seen in the context of local understandings of childhood, expectations about children's roles and responsibilities, and the contributions that children make to their families. A complex mix of factors, including poverty, gender, caste, ethnicity, parental ill-health, and attitudes to work, appear to intersect to explain why children work. Far from passively being put to work, some children make active decisions to help the family situation by working.

Qualitative research with children

Using an analysis of survey data from Young Lives, two rural sites were identified as having relatively high rates of older cohort children involved in agricultural wage labour. The study explored the following questions. First, what are children's understandings of 'risks' at work? Second, what are children's experiences of managing risks they face at work?

Children engaged in agricultural work, either on their own farms or for wages, were invited to take part. Methods were adapted from Woodhead's (1999) study of the psychosocial consequences of children's work in six countries, using activities and questions designed to elicit how children perceive their work. Interviews were conducted with children, exploring their experiences of work, time use and responsibilities, managing risks, and what they perceived might be the risks in stopping children from working. The interviews included 'body mapping', a visual method using a large sheet of paper with the outline of a body that enables children to identify parts of the body affected by work. This is a recognised method for eliciting people's experiences of health and harm (see Clacherty 2009; Cornwall 1992). Group discussions explored children's daily lives, time use, good and bad aspects of work, school, combining work and school, which work is best, and at what age childhood begins and ends in relation to different tasks. Group discussions with caregivers focused on community norms affecting children's activities, risks to children and how they manage them, and variations according to gender, age, birth order, and caste. Children and adults gave consent. Fieldwork was conducted in Spring 2011, and fieldwork teams spent about ten days in each site.

We use pseudonyms for the two sites. Avanti is in Rayalaseema, where the main crops are paddy, sunflower, sesame, millet, cotton, sweet-lime, lime, and mango. Children work most frequently in the sweet-lime groves, with work in lemon groves being the next most common. Twenty-one children aged 15–16 years were interviewed, 11 girls and 10 boys. Compulsory education lasts until age 14 in India, and most of them were in full-time work, although three boys were combining school and work. Most were from Scheduled Castes or Backward Castes.[1] Bhakrapet is in Telangana region of the state. Crops are paddy, chillies, and cotton. Children are involved in herding livestock, involving stays of two to three months in the forest while they graze cows and sheep. Children also work in chilli and cotton farming (not for seed). Twenty-one children were interviewed, 10 boys and 11 girls (11 were from Scheduled Castes, five from Backward Castes, and five from Scheduled Tribes). Three group activities were held with boys, three with girls, and individual interview was held with *sarpanch* (elected head of the village-level self-governing council), and two collective interviews were held with caregivers.

Why children work

There are a number of factors that explain why children work. First, poverty or specific household situations may leave children with little choice about work. They engage in work in different crops, and finding work is important to them. Second, according to children, parents think that children are ready to work as they turn 12 years old. Once they enter work, they continue to work, and this gradually becomes their occupation. This work prepares them for the future. Children from not-so-poor households who have left school take up work on the farms on the days that they like to work, and have some control over their earnings. These children prefer to work with groups of friends and look for fun and enjoyment while working, despite the physical demands of work. Children describe using their earnings to buy clothes, footwear, CDs, mobile phone items, jewellery, and other consumer goods. Other benefits of working included enhancing marriage prospects, having something to do, providing a source of pride, having fun, and a way of learning skills for the future.

For many children, their experiences of school play a role: some described how they dislike school, for many reasons, one of the most frequently cited being that teachers beat them if they did not do well in their studies. One boy in Avanti described how he did badly at school:

> *I am poor in studies. It is a waste ... if I study. If I continued to study ... doing hard work will be later on ... so I did not study ... I think work is better for me.*

Another boy, in Bhakrapet, described how he had completed his schooling, and started working on the farm:

> *What else can I do, sitting at home all the time, the day long? It is terribly boring and anyway we do need some money for the household. I started to work as I felt there was*

> *a need for some more money ... No one asked me to work and they wanted me to remain at home but I felt very bad remaining at home due to boredom. So I thought it would be better to work and so I started to go and work.*

Other boys described how they chose to work. One boy in Avanti described how:

> *We ... feel like going to work ... When we see all the others going to work, naturally we too would feel like going to work. No one would like to stay alone at home ... if we go and work, it will be of immense help to us ... if we go work, and are able to give our parents some money, they too would reciprocate by giving us say some 20 to 30 rupees for our expenses. What is the point of staying at home without working? It is not of any use and why should one stay at home without working?*

Working is fundamental for family relationships. One boy described how he undertook a great deal of work on the family farm, because his father was living in Kuwait. He liked the work, and work was a source of pride. Another boy said, '*I am earning and giving to my parents, so I feel proud of that*'. Work is also important for children's relationships with others in the community, and children's reputations. Work is part of being a 'good' child. One boy recounted how his mother had been told by others in the community:

> *'Look, you are a blessed one. You are being looked after by your son and there is no need for you to work. He is not only earning but also taking care of you.' I felt very happy. I want to get a good name, still want to work hard and do better things.*

Another boy described how working provided money that gave him status. He recounted how before he worked,

> *nobody used to come to me; now since I have money, everybody comes, so I understood that everything will be there with the person who has money.*

Work was seen as beneficial as a form of education in itself, because children learn work skills that they will need when they marry. For girls, prospective in-laws '*will certainly ask what we have learned in our parents' house*'. As one girl said:

> *After they get married and go, when there is more work then the work learned here will come in very useful, and if they go there without these skills, it would be very disadvantageous. So it is better to learn as they will be equipped for the future.*

Additionally, work teaches children skills that are useful in general. One girl said, '*When we grow up, we can do the work*'. One boy commented about his work: '*Since we have no education, this is useful*'. When asked what he had learnt by working from an early age, one boy commented,

> *I can do things on my own. I need not depend on others. I feel like that.*

In discussion, children in Bhakrapet mentioned that the suitable age for children to start work is 17 to 20 years. However, children talked about how occasional work was useful as training for full-time work and suggested that they should start earlier than that, so that they would be prepared for work by the time they are 18 or 20. '*Learning work should start early, or else the learning will never take place, because it is difficult to learn as adults*'. On the whole, however, children hoped or planned to move out of agricultural work altogether. At present, they felt that agriculture is not viable for survival, it was not profitable because of crop diseases, and it involves hard work in the hot sun.

When children work, they also socialise with friends, and have fun. One girl described how she preferred working on other people's fields, and she regretted working on her own on her family land:

Because I didn't get the opportunity of being with friends and being happy talking [about] different things and enjoying [myself]. When we are working on our land we are alone, whereas in the wage type of work, we get company... when I am in my own field, my eyes are always on the neighbouring fields ...

One boy talked about how regular work made a person stronger:

In case one works regularly and consistently and works well, that person will not suffer from anything like fever or other ailments. He will not have any complications at all ... He will be hale and healthy ... if he goes to work as usual every day ... If one remains at home without doing anything he will turn into a vegetable in the course of time. He will become too weak and suffer from fever and other complications.

Children discussed what would happen if they were prohibited from working. For several children, being prevented from working would mean that they would have no food. One girl commented:

We will not have rice, vegetables, and if we fall sick, they will not be able to take care of us. In every way, they face problems.

One boy commented:

I cannot imagine a life without working. Work means everything to me. Unless I work, we cannot run our house ... We need to buy food to run our home ... and I have to work to raise money to buy all these.

Another boy, in Bhakrapet, commented, '*If I completely stopped working, I would be as good as dead, there will not be any life left within me*'.

On the other hand, not working or attending school means that some children '*just loaf around doing nothing ... they are drinking and smoking beedis. They are very indulgent*'. Not working was perceived as reducing one's marriage prospects. One boy commented:

What would happen to me if not working? My whole future is already spoiled. It would have been ruined even more. Studies are no more. I used to tend goats, even that is lost. Now if I lose [work] also, I would never get married.

Experiences of agricultural work

Children undertake different operations at each of the stages of the life cycle of the various crops. Operations are clearly demarcated by gender, and a general agreement that the tasks done by boys and men are more physically risky and difficult. For example, in paddy, operations include preparing the field (ploughing), sowing, transplanting, weeding, spraying of fertilisers and pesticides, watering, harvesting, bundling, thrashing, collecting grain, and marketing. Of these tasks, men and boys are responsible for ploughing, watering, and thrashing, while women and girls are engaged in transplanting, weeding, harvesting, and bundling. The gender-based division of labour is found in almost all crops in both sites. Children's work on farms inevitably involves the risk of injuries and physical strain. Paddy is perceived as the most unpleasant, because it involves standing in water for planting, working in hot sun, working for low wages, and children felt they had to work in order to please the landlord. Table 1 describes the different operations and

Table 1: Work activities, hazards, and responses, identified by young people

Work activity	***Part of body affected***	***Harm at work, and responses***
Avanti: Paddy –Ploughing/ preparation of land for planting of seedlings	Legs – severe muscle pain	Sad about doing this kind of work
Paddy –Trenching and forming irrigation canals	Moderate muscle pain in neck, hands and shoulders. Medium injury to legs or toes.	Not able to work speedily the next day, can't walk properly. Then they lose their wages for a further few days.
Cutting crop of paddy during harvest	Cuts to fingers while cutting	Fear of the pain
Picking lemons and oranges	Scratches from thorns on the back of hands, neck pain, and headaches	Feeling pain if scratched, but 'it is a pleasure to work with the other children'
Carrying the picked lemons – 75kg bags	Backache	It is well paid, and there was no other option to earn
Spraying pesticide and spreading fertilizer	Severe stomach upset, burning sensation, loss of appetite, uneasiness in the body for up to one or two days	Children are not interested in doing this job, even though it is a half-day's work and well-paid. Some children reported being prevented from spraying by parents.
Picking sunflowers	Mild irritation and itching to the hands	Wear plastic gloves
Mixing sand and cement in construction work	Severe cracking and burning of hands and legs	Wearing gloves or polythene covers. Applying oil after reaching home.
Carrying bricks	Severe shoulder pain	Carrying smaller loads
Rubbing off old paint and whitewash	Shoulders ache, hands hurt, sometimes it falls into the eyes	
Roaming where the livestock goes	Continuous standing – legs ache	Feeling strained, not able to focus on studies after returning home
Bhakrapet: Clearing land for paddy, weeding	Scorpion stings, snakebites very common. Hands, fingers, feet affected. Also, thorns and stones.	Sadness. Cured by juice of leaves prepared by a person in the village. Treatment used with *mantras.*
Ploughing	Blisters	Tiring, hard work. Rest afterwards. Ideally, men do this work, but if fathers are absent, older boys undertake the work
Herding livestock	Risk of attack by bears, other wild animals	Take dogs, axes, sleep together in groups
Sowing	Involves bending down, causing muscle aches and pains.	Rest
Irrigation work	Electric shocks Night time working	Being careful Rest during the day

Table 1: Continued

Spraying pesticide	Vomiting and diarrhoea. Sometimes feel giddy and faint.	Recently one boy was taken to private hospital and was given glucose.
Picking chillies	Need to work in hot sun. Body, face risk sunburn. Hands burn. Scorpion stings and snake bites common. Hot and pungent, and huge crops.	Wear long-sleeved shirts like men, bear the pain and do the work. Cover head with cloth, take a towel to mop sweat, and do the work carefully. Avoid the work, if allergic to chillies.
Transplanting paddy	Feet and hands affected as have to be in water much of the day. Snakebites and scorpion stings common.	Home-based treatment, medicines from pharmacy. If it is serious, like a snake bite, they go to hospital.

parts of the body that are affected, and how children felt, and responded to risks. The content of the table was generated by children in group discussions, and some responses included risks at work beyond agriculture. Additional data are included from interviews. The intention is not to suggest that all children take the protective measures mentioned, but to illustrate their knowledge of what these might be.

Harm children reported fell into three broad themes. The first includes physical hazards, such as cuts, scratches, and wounds while harvesting, cutting firewood, and ploughing; bruises from thorns and twigs while climbing trees and picking fruits; irritation from picking sunflowers; injuries from falling from the plough or from trees; snakebites and scorpion stings; and attacks on herders by animals. Another risk was of electric shocks from water pumps, and from electric fencing. The second category comprises health-related problems like itching, skin irritations, stomach ache, muscle and joint pains, headaches as a result of working in the sun, stomach upsets and nausea from pesticide exposure. In Bhakrapet, for example, boys described getting up at 3–4 am to irrigate paddy fields, because the electricity supply is intermittent. This made them tired during the day. The third are psychosocial effects, such as disliking work or not wanting to work, feeling looked down upon by peers for not attending school, being teased for working by other children, and fearfulness and anxiety caused by work. Boys undertaking irrigation work at night described being afraid of evil spirits and ghosts. One boy described going to a traditional healer to ward off spirits, but said he couldn't work for two months. Children also mentioned the adverse effects of working on educational attainment, as one boy said, '*Our education will be spoiled ... the children who go regularly to school score more marks than we do*'.

Precautions taken against threats at work

Children described numerous ways in which they tried to protect themselves at work. To avoid injuries while chopping wood, they undertook the task slowly and carefully. When moving stones, they spoke of carrying a little at a time to avoid injuries. When herding cattle, they carry axes, in case something attacks them, and they sleep together in groups, with dogs.

> *There would be ten of us sleeping together, and we can fight anything that intrudes ... we would have dogs with us ... When in danger, they would start barking.*

They described wearing long trousers and long-sleeved shirts when harvesting paddy to avoid scratches. If working with pesticides, children mentioned wearing a 'helmet' (protective face mask), that is, if it can be afforded. Children reported that not many people purchase them because they cost 400 rupees. Alternatively, they described tying or holding a cloth over the face but being aware that the pesticides may still affect their eyes. One boy said,

'I cover my mouth and nostrils with a mask, the way they cover themselves in hospitals'.

Pesticide spraying is well-paid compared to other tasks, so it is an attractive but potentially dangerous task.

Children also described ways of avoiding injury in work other than agriculture that takes place in rural areas, probably helping or replacing parents in work undertaken as part of the National Rural Employment Guarantee Scheme. One boy described avoiding blisters while working with cement in construction work: *'I do not touch the mixed concrete with my bare hands*'. Another boy described how his mother would not let him do the '*government jobs where they uproot trees ... My mother says no ... I might get hurt there... so she does not want me going there*'. Children also mentioned wearing *chappals* (shoes) to protect themselves when walking or working on stony ground:

> *We wear those rubber soled shoes ... the ones made with tyre material ... everybody wears those only while working. So there is not much problem.*

Some children described avoiding certain tasks. The most universally disliked work was related to paddy. One boy said he did not crop paddy, for fear of what might have happened (no children reported actual loss of fingers):

> *It's not because I cannot do that job. If we do that job, our fingers are likely to get injured ... the sickle slips off our hands and lands on our fingers, and one of them gets chopped off ... I had been once to this job, and my fingers were injured. They were not completely chopped off .... From that time, I stopped doing that job.*

A girl mentioned that she did not harvest paddy, because of the itching caused by the straw. '*That's why I don't go... I don't like it ... we get it done by using wage labour*'.

Another boy talked about being prohibited from spraying pesticides altogether.

> *Pesticides are very strong and pungent, so they [parents] will not let me do that. It is difficult, they will not allow me to do that work. They say, 'Look, don't do that. It is difficult and you cannot do that. In case you become unconscious ... we have no money to get you cured. Don't do that.' That is what they said and so I don't do that job.*

One girl described collecting oranges, but not lemons, because of the thorns, and not cutting paddy. Another boy mentioned that his mother only lets him do simple tasks, *'such as planting, picking lemons ... especially lemon picking because it is just sitting under a tree and picking out lemons'*.

Several examples show how children actively engage with the difficulties they face at work. This is not to suggest that these are acceptable risks, but to illustrate that they do have knowledge, experience, agency, and they are not simply passive victims in all this. For example, children explained that they were more likely to be injured at work if they lacked experience. Injuries happen mainly when they were new to the work, and these reduce as they get used to the tasks. One girl mentioned that the work *caused wounds to legs ... spade struck legs ... when [we] go to work with half-knowledge about the work ... things like that happen. Not now ... I do better now ... now I do it carefully ... not much happens now.*

Several incidents mentioned by children showed their initiative in dealing with their problems. When they were injured, children mentioned seeing a doctor, mostly on their own, or buying a pain killer or a fever tablet across the counter. While herding livestock in the forests, two boys in Bhakrapet reported fear of being attacked by bears, and described how the night stays in the forest are scary, but their strategy was to take dogs with them, and axes to ward off attacks.

Linking micro-level data from children to international aims in child protection

Thus, children describe a range of benefits of working, as well as the risks of not working. They often perceive work as something they choose to do rather than being forced into, and work may be seen as a useful alternative to schooling, even though it can involve unpleasant drudgery. Children are involved in a range of agricultural activities, and they are aware of many of the hazards. While some children avoid tasks that are particularly hazardous and unpleasant, not all do. Few children suggest that avoiding work altogether is a solution to their problems. The decisions made by children and their families need to be taken into account if intervention is to be genuinely protective. What are the implications of this for thinking about hazardous labour, given that children (and their families) appear to be aware of the hazards they face and have some strategies for dealing with them?

In contrast to trying to eradicate the worst forms through the elimination of all child labour, pragmatic approaches grounded in children's experiences (Bourdillon *et al.* 2010: 194–202), but also advocated by occupational safety and health experts (Fassa *et al.* 2010), may offer strategies for making work safer, not just for children, but also for adults. However, for pragmatic approaches to be accepted, an ideological shift would be needed among policy actors at international, national and state levels – away from the powerful but unrealistic rhetoric associated with child labour. Children work on family farms in a number of activities that involve risk of harm in the most developed economies, and it has not proved possible to 'eradicate' this phenomenon entirely. Effective regulation, particularly in relation to specific risks such as pesticide use, could be developed, though in some instances children already appear to be prohibited from this informally (by parents).

There are also broader questions about the sustainability of farming, and the acquisition of knowledge and skills about agriculture, if children are forced to withdraw from work. Children do not seem to value agricultural work, although realistically, this is where the future is likely to lie for many of them. However, if children see no future in farming, this has implications for the sustainability of subsistence and market agriculture in developing countries. Enormous structural changes are needed before children's work is no longer required. In the meantime, in the current transition phase of agricultural development, it would be more respectful of children's dignity at work to recognise their activities, instead of devaluing their work and constructing it as a problem. Realistic questions also need to be asked about the proportion of the population that is likely to be able to escape farming through skills acquired at school.

Finally, we question the value of focusing solely on 'hazardous child labour'. It has long been acknowledged in rural sociology that in poor communities, traditional agriculture cannot be fully disentangled from practices of everyday life. Children's work is one resource among many that poor households draw upon in strategies to maintain their livelihoods. It is perhaps more helpful to focus on all household members and their livelihoods, instead of singling out children for study and intervention. Where daily survival is challenging, work is fundamental and cannot be considered separately from other aspects of daily life, and therefore the poverty behind it. Work is also a crucial aspect of children's social relationships and cannot be separated from inter-relationships within families and communities. Further, many health hazards in poor rural communities are not related to work, but to living conditions in general. Consideration must be given to what will be lost if children are removed from agricultural work altogether. Work may be integral to education, in the broadest sense, and may develop children's potential in the situations in which they live.

The common focus on either eliminating children's work altogether, or importing health and safety standards, is often based on the paternalistic assumptions that workers in developing countries (adults and children) do not know how to protect themselves. Data gathered from children about their experiences have shown that communities may develop their own methods of minimising harm at work, appropriately adapted to local circumstances. Children point to specific tasks, related to particular crops, and much can be learnt from their experiences. However, further investigation is needed, because we cannot assume that poor rural communities are fully aware of all hazards that affect both children and adults alike, nor can we burden them with full responsibility for developing their own safer strategies. Research is therefore required to identify specific areas where safety needs to be improved, working alongside community norms instead of against them.

Acknowledgements

The authors thank the children, parents, and others who participated in Young Lives research and the qualitative study reported here, the research team in Tirupati, Daniel Vujcich for valuable research assistance, and Gina Crivello, Dr Vijay Kumar, and an anonymous reviewer for helpful comments on an earlier draft of this paper. Young Lives is funded by Department for International Development (DFID). The qualitative study was funded by the Oak Foundation. The views expressed here are those of the authors and not those of DFID, Oak Foundation, Young Lives, or University of Oxford.

Note

1. The caste system is still extremely important in India in various spheres, not least politically. Scheduled Castes (SCs) are the lowest in the traditional caste structure and were earlier considered to be 'untouchables'. In rural Andhra Pradesh, SC colonies are located away from the main villages. SCs have been subjected to discrimination for years and had no access to basic services, including education. National legislation aims to prohibit 'untouchability' and discrimination. Backward castes or classes (BCs) are people belonging to a group of castes who are considered to be 'backward' in view of the low level in the caste structure.

References

Abebe, T. and S. Bessell (2011) 'Dominant discourses, debates and silences on child labour in Africa and Asia', *Third World Quarterly* 32 (4): 765–86.

Bourdillon, M., D. Levison, W. Myers, and B. White (2010) *Rights and Wrongs of Children's Work*, Camden, NJ: Rutgers University Press.

Clacherty, G. (2009) *Hard Work, Long Hours and Little Pay. Research with Children Working on Tobacco Farms in Malawi*, Lilongwe, Malawi: Plan Malawi.

Cornwall, A. (1992) 'Body mapping in health RRA/PRA', *RRA Notes* 16: 69–76.

Fassa, A., D. Parker, and T. Scanlon (eds.) (2010) *Child Labour. A Public Health Perspective*, Oxford, UK: Oxford University Press.

Gupta, D. (2005) 'Caste and politics: identity over system', *Annual Review of Anthropology* 34: 409–27.

ILO (2010) 'Road Map for achieving the elimination of the Worst Forms of Child Labour by 2016', available at http://www.ilocarib.org.tt/images/stories/contenido/pdf/childLabour/ROADMAP%202016.pdf (retrieved 1 October 2011).

ILO (2011) *Children in Hazardous Work. What We Know. What We Need to Do*, Geneva, Switzerland: ILO/IPEC.

Levison, D. and M. Murray-Close (2005) 'Challenges in determining how child work affects child health', *Public Health Reports* 120: 614–20.

Morrow, V. and U. Vennam (2010) 'Combining work and school: the dynamics of girls' involvement in agricultural work in Andhra Pradesh, India', *Children & Society* 24: 304–14.

Ramanathan, U. (2000) 'The public policy problem: child labour and the law in India', in B. Schlemmer (ed.) *The Exploited Child*, London, UK: Zed Books.

Shirol, S. V. (2000) *Child Labour: A Rural Reality*, New Delhi, India: Classical.

Subbaraman, S. and H. von Witzke (2007) 'An Overview of Child Labour in India', in G. Herath and K. Sharma (eds.) *Child Labour in South Asia*, Aldershot, UK: Ashgate.

Woodhead, M. (1999) 'Combating child labour: listen to what the children say', *Childhood* 6 (1): 27–49.

'Risky lives': risk and protection for children growing-up in poverty

Kirrily Pells

Child protection has focused on responding to interpersonal violence and abuse. This approach can detach children from the broader socio-economic and political structures which shape their life chances, by concentrating on the symptoms of risk rather than the underlying causes. Drawing on the Young Lives study of childhood poverty, this paper argues that poverty and inequalities are at the heart of childhood risk, shaping which children are at risk and access to sources of protection and therefore to children's life chances. In order to protect children better, the child protection field needs to engage with the structural nature of risk.

Des 'vies risquées': risques et protection pour les enfants qui grandissent dans des conditions de pauvreté
La protection de l'enfance s'est concentrée sur la réponse aux violences et aux abus interpersonnels. Cette approche peut éloigner les enfants des structures socio-économiques et politiques plus larges qui façonnent leurs chances de vie, en se concentrant sur les symptômes des risques plutôt que sur leurs causes sous-jacentes. En utilisant l'étude Young Lives de la pauvreté parmi les enfants, cet article soutient que la pauvreté et les inégalités sont au cœur même des risques pour les enfants, et qu'elles déterminent quels sont les enfants qui courent des risques et leur accès aux sources de protection et donc aux chances de vie. Pour mieux protéger les enfants, le secteur de la protection de l'enfance doit mieux tenir compte de la nature structurée des risques.

'Vidas arriscadas': risco e proteção para crianças que estão crescendo na pobreza
A proteção infantil tem se concentrado na resposta à violência e ao abuso interpessoal. Essa abordagem pode separar as crianças das estruturas sócio-econômicas e políticas mais amplas que influenciam suas chances na vida ao concentrar-se nos sintomas do risco em vez das causas subjacentes. Baseando-se no estudo de Young Lives sobre a pobreza na infância, este artigo argumenta que a pobreza e as desigualdades estão no centro do risco da infância, influenciando tanto quais crianças estão em risco quanto o acesso a fontes de proteção, afetando desta forma as chances nas vidas das crianças. Para proteger melhor as crianças, a área de proteção infantil precisa engajar-se com a natureza abrangente do risco.

'Vidas en riesgo': riesgo y protección para los niños que crecen en condiciones de pobreza
La protección de la niñez se ha centrado en cómo responder ante la violencia o el abuso interpersonal. Esta propuesta puede desvincular a la niñez de su contexto socioeconómico y político general, el cual determina las oportunidades que tendrá en la vida, ya que la propuesta se enfoca en los síntomas del riesgo y pasa por alto las causas estructurales. Basado en una investiga-

ción de Vidas Jóvenes sobre niños y niñas que viven en la pobreza, este ensayo sostiene que la pobreza y las desigualdades constituyen las razones principales del riesgo que enfrenta la niñez, ya que determinan quiénes están en riesgo y quiénes tienen acceso a mecanismos de protección que aumentan sus posibilidades de sobrevivir. Para que los niños y niñas estén realmente más protegidos, necesitan analizarse más a fondo las causas estructurales del riesgo.

Introduction

Child protection aims to '*prevent, respond and resolve the abuse, neglect, exploitation and violence experienced by children*'. While acknowledging that the '*root causes of child protection failures*' include '*chronic poverty, insecurity, power imbalances and harmful traditional attitudes and behaviours*', in practice focus has been predominately on responding to interpersonal violence, abuse, and neglect experienced by children.[1] Such an approach encourages a focus on certain 'categories' of children, such as 'orphans and vulnerable children', perceived to be most exposed to risk, as well as extreme situations such as armed conflict, prostitution, and hazardous child labour. This approach also detaches children from the broader socio-economic and political structures that shape their life chances and trajectories, by concentrating on the symptoms of risk rather than the underlying causes of poverty and inequality.

Chronic poverty is a significant risk factor for poorer life chances, as well as increased the chances of exposure to other risks, such as drought, illness, poor nutrition, harmful forms of labour, and poor quality, or lack of, services. Across the life course, these risks may have different impacts, as well as having cumulative effects lasting into adulthood and reinforcing the transmission of poverty and inequalities to the next generation (Boyden and Cooper 2009; UNICEF 2010). At the same time exposure to risk does not always result in negative outcomes. Among children growing up in adverse circumstances some fare better than others (Luthar *et al.* 2000; Werner and Smith 2001). Understanding the nature of risks faced by children, their households, and communities, as well as sources of support and protection, is therefore essential in protecting children and improving their life chances.

Drawing on quantitative and qualitative data from the Young Lives study of international childhood poverty in Andhra Pradesh (India), Ethiopia, Peru, and Vietnam, this paper argues that poverty and inequalities are at the heart of childhood risk, shaping which children are at risk, access to sources of protection, and therefore children's life chances. The paper begins with a discussion of risk, resilience, and protection before providing an overview of the Young Lives study. This is followed by three key sections. The first examines how disparities in children's life chances widen over the life course, with risks being cumulative and compounded, which, as discussed in the second section, frequently disadvantage the same children. The third section explores the complex interconnections between risk, protective processes, and children's life chances, through the lens of family illness and death. The conclusions are that to protect children better, the child protection field needs to engage with the structured nature of risk faced by children, their households, and communities, and to build on sources of protection within them.

Risk, resilience and protection

The concepts of risk and resilience have become increasingly popular in understanding the impact of negative circumstances on children's development, and factors that enable children to '*recover from, adapt to and/or remain strong in the face of adversity*' (Boyden and Cooper 2009: 289). While risk is understood as factors which increase the probability of negative outcomes, there is much less conceptual and methodological clarity over resilience. Consequently, this paper uses the term protective processes, suggesting ongoing actions that may moderate and mitigate the impact of risk on children and their families.

Research framing child protection issues in terms of risk, resilience and protective factors often tends to emphasise internal psychological processes over broader structural factors and social processes (Boyden and Cooper 2009; Crivello and Boyden 2011; for an exception see Werner and Smith 2001). Individualistic approaches target symptoms rather than the root causes of the problem and '*de-politicises the project of poverty reduction*' (Boyden and Cooper 2009: 295; Seccombe 2002). Moreover, it tends to result in a focus on categories of children 'at risk', such as orphans and vulnerable children, and crisis situations such as humanitarian disasters, rather than the underlying structural causes of risk. The subjective perceptions and collective understandings of those affected also tend to be overlooked in favour of 'objective' standards of negative and positive developmental pathways (Crivello and Boyden 2011).

Some factors may simultaneously represent risk and protection, implying that they can either contribute to or undermine life chances. For example, children may leave school in order to work to support the household because of the death of a caregiver. This may be a risk factor for the longer-term chances of the young person obtaining a higher-paid job and potentially may be damaging to their health and sense of well-being. At the same time this strategy stops the household from either disintegrating, starving, or having to take out a loan at a high interest rate, and the child may develop more vocational skills and a sense of achievement by supporting the family. This is part of a series of trade-offs that families living in chronic poverty are forced to make (Kabeer 2000), especially in coping with adverse events, as emerges in the following sections.

Young Lives and methodology

Young Lives is a longitudinal study of childhood poverty which has been collecting both survey and qualitative data on around 12,000 children since 2002 in four countries (Ethiopia, the Indian state of Andhra Pradesh, Peru, and Vietnam). This paper focuses on the trajectories of an older cohort of children of around 1,000 children in each country (slightly fewer in Peru) born in 1994 and 1995. Household and child-level data were collected on this group when they were eight, 12, and 15 years old. In-depth qualitative information was collected on 25 older cohort and 25 younger cohort children in each country in 2007, 2008 and 2010/11. All the names of Young Lives children used in this paper are pseudonyms.

The study focuses on children growing up in poorer populations.[2] The data therefore are not nationally representative; nor should they be compared simplistically between the Young Lives countries, since the samples of children are drawn differently. Such panel data can, however, highlight how change over time is affecting children and what similar (or different) processes are taking place in each of the countries, as well as disparities within the samples.

This paper uses three principal methods. Descriptive statistics are used to explore changes in children's life chances between the three survey rounds, disaggregated according to socio-demographic characteristics of the child and the household. Regression analysis is used to examine the impact of a series of potential risk and protective factors reported in 2006 upon school enrolment, cognitive achievement, self-rated health, BMI(Body Mass Index)-for-age, height-for-age, and subjective well-being measured in 2009 of children in the older cohorts across all four countries. This includes the socio-demographic characteristics of the child and household, risk factors (including economic and environmental shocks, illness and death of a family member) and protective processes (including household membership of groups and level of support, and access to services). Although this analysis does not necessarily prove causality it can be used to illustrate factors which are associated with one another. Findings are supported further by interviews with children, their caregivers and community members, from Ethiopia and Andhra Pradesh.[3]

Recognising the methodological challenges in capturing the complexity of interactions between multiple potential risk and protective factors, this paper is cautious in attributing exact causation. Instead, it identifies channels through which factors may have a negative effect on children's life chances, such as lower levels of cognitive development or poor health, or may assist children and families in responding to risk.

Cumulative and compounded risk

For children growing up in poverty risk is '*an integral part of everyday "ordinary" lives in which the young negotiate multiple, interacting challenges*' (Crivello and Boyden 2011: 1). Risk can be cumulative and compounded, with factors interacting with one another and thereby having an increasingly negative impact on children's life chances over time. This is explored in the following section, which examines children's trajectories through schooling and outcomes in formal learning.

Children's experiences of school and formal learning outcomes are shaped not only by their socio-economic status and household situation, but also by factors within school. These can include the quality of education received, challenges in the school environment such as lack of gender-segregated toilets or forms of corporal punishment, as well as differing attitudes and expectations of teachers for boys and girls (Pells 2011). Factors internal and external to the school environment may therefore hinder progression through school, increasing the risk of grade repetition and of children leaving school before they complete a full course of primary education.

This is illustrated by the case of Mohan, who was 15 years old at the time of data collection and from a Backward Caste in rural Andhra Pradesh. He stopped going to school after the seventh grade as his father was sick. He did not return to school, even though his parents encouraged him to do so. He said '*everyone will be new and I thought it will be awkward if I go back so I never went*'. Previously he used to be beaten by teachers as he was absent frequently. Mohan also explains that in the future he will be responsible for supporting his parents as well as his own family when he marries, and says staying in school is '*risky*'. He believes it is important to learn skills like his father, who is a mason, so that he can earn money to have a good life. Mohan lists the range of skills he has learnt from his father and older cousins including ploughing, driving a bulldozer, driving a bull cart, mending the tractor, building walls, and plastering. It illustrates the series of trade-offs made by families living in chronic poverty and underscores the need to consider schooling in the broader context of economic development and whether it is equipping children like Mohan with the skills that they need in order to enter the local labour market.

Within Young Lives study countries, primary schooling has increased, particularly in areas and among groups with initially lower enrolment levels. Yet rising enrolment rates do not fully capture children's pathways through schooling and how and why these may vary between different children. Despite the different stages of economic development and socio-cultural contexts of the four countries, there is a consistent pattern of disparities, which widen over time. As the children become older, differences in school enrolment and cognitive achievement between non-poor children and poor children, between children living in urban and rural areas (illustrated by Figure 1), and between groups with different social status because of ethnicity or caste, increase.

Across all four countries household wealth (a composite measure which includes consumer durables, quality of housing, and access to services) is a predictor of school enrolment. As the children grow older, those from the poorest households are least likely to be enrolled and have lower levels of cognitive achievement. For example, in Vietnam the percentage point difference in school enrolment level between poor children and the rest more than tripled between the ages of 12 and 15. Controlling for other factors, the household position on the wealth index is also associated with children's cognitive achievement. Cognitive achievement is measured for all children, not just those in school, and the score used here combines two measures, one testing receptive vocabulary and the other testing maths.[4] The wealth index is constructed on a scale with 0 (minimum) and 1 (maximum). Across all four countries children who live in wealthier households performed better in the cognitive achievement tests. Each point on the wealth index accounted for 1.6 points on a five-point cognitive score of children in Andhra Pradesh, 1.7 points in Peru, 2 points in Vietnam and 2.2 points in Ethiopia.

In both Vietnam and Peru there is a large gap between children in the poorest quintile and the rest in terms of cognitive achievement. Further analysis suggests that this may be due to the marginalised status of ethnic minorities within the countries. This is illustrated by Figure 2, which shows that in Vietnam children from ethnic minorities are over three times more likely to be in the lowest category of cognitive achievement than their Kinh (majority ethnic group) peers. Moreover, while the numbers of Kinh children in the bottom category declines between 2006 and 2009, there is no change among children from ethnic minorities.[5]

A similar situation exists in Peru, where only 10 per cent of children from ethnic minorities attend bilingual schools, the rest receive education in Spanish. Young Lives research also found that half the teachers in bilingual schools in southern Peru could not speak the local indigenous language and that bilingual materials were not being used (Cueto, Guerrero, *et al.* 2009). These disparities are indicative of an 'inverse care law' whereby poorer and more marginalised children often attend schools with fewer resources and receive a poorer-quality education (UNESCO 2010: 23).

While important, differentials between girls and boys are much smaller than between children living in urban areas and those living in rural areas (Pells 2011). This is a situation reflected across the countries. However, important trends are emerging as the children grow older which have implications for the reproduction of inequalities and children's transition beyond school to liveli-

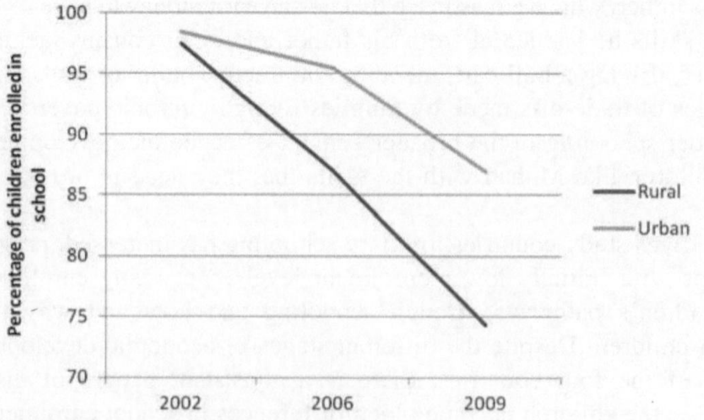

Figure 1: Widening gaps in school enrolment, Andhra Pradesh Older Cohort, 2002–09 (n. = 1008)

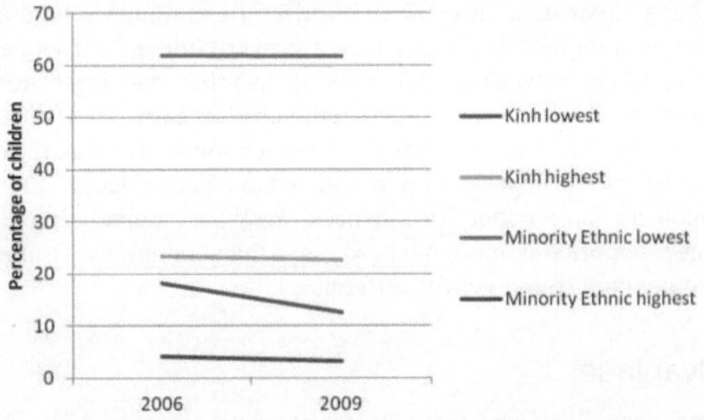

Figure 2: Children in the lowest and highest categories of cognitive achievement in Vietnam, according to ethnicity, 2006 (n. = 925) and 2009 (n. = 961)

hoods, the labour market, and the establishment of households of their own. Though within the Young Lives samples there was parity in the numbers of boys and girls going to primary school, national data show that this has been achieved in Vietnam and Peru, but that gender parity does not yet exist at primary level in Ethiopia and India (UNICEF 2010). By 2009 differences between boys and girls have become clearer yet, contrary to the dominant narrative of girls being consistently disadvantaged (Levine *et al.* 2008), within the Young Lives sample it is not always girls who fare less well. In Peru, Vietnam, and Ethiopia more boys than girls have left school by the age of 15 years. This could be due to the higher wage-earning potential of boys, or boys performing less well in exams (UNESCO 2010: 67). In addition girls may be able to combine tasks at home with studying, but boys may find it harder to combine school with work, given the different types of work undertaken according to gender (Pells 2011).

In Peru and Ethiopia boys are increasingly outperforming girls in cognitive achievement tests over time. Conversely, in Vietnam the number of girls in the highest quintile of performance is increasing, while the number of boys is decreasing. In Andhra Pradesh there is a consistent pattern of girls being disadvantaged, with boys 1.8 times more likely to be enrolled by 2009 and performing better in cognitive achievement tests. This may become increasingly significant for future generations, given the importance of maternal education for Young Lives children.

Studies have shown maternal education to be a significant factor in breaking the intergenerational transmission of poverty, as girls who stay in school for longer marry and have children later, so decreasing the risk of maternal and child mortality and morbidity (Levine *et al.* 2008; UNICEF 2010: 4). Children with mothers with no or little education tend to have lower levels of cognitive achievement, progress through school more slowly and leave school earlier.

More analysis is required to understand the relationship between maternal education and household socio-economic status. Although maternal education is a fairly robust proxy for other household factors, it is likely that it reflects rather than wholly drives household disadvantage. In other words, households with a history of disadvantage also tend to have mothers with lower levels of education. However, regression analysis that controlled for a series of variables including household wealth, found that children with more educated mothers tend to have higher cognitive test scores in all four countries. Each year of maternal education correlated with slight increases on the cognitive achievement scale. This may be because more highly educated mothers are better able to support children's formal learning and have higher bargaining power to allocate resources within the household to educational and healthy development (Levine *et al.* 2008: 20).

In summary, Young Lives data on children's trajectories through school and outcomes in formal learning demonstrate that disparities increase over children's lives between non-poor children and poor children, between children living in urban and rural areas, and between groups with different social status because of ethnicity or caste. The differences between boys and girls also increase over time, although at a much lower rate than the other disparities. However, maternal education appears as a significant mediating factor: children with more highly-educated mothers have better life chances. Risks are cumulative and compounded across the life course, but often in more nuanced ways than commonly assumed, as illustrated by the complex picture concerning gender differences.

Multiple disadvantages

Chronic poverty and inequalities repeatedly disadvantage the same children, who are at a higher risk of having poorer outcomes in multiple areas of education, health and subjective well-being. This section brings together the education-related areas of the previous section with indicators of healthy development and nutrition. Inadequate nutrition, including insufficient food, poor micronutrient intake, and disease at an early age can result in stunted growth. Young Lives data have demonstrated a number of consequences of stunting including poorer cognitive development in Vietnam (Le Thuc 2009) and Peru (Sanchez 2009). Stunting also impacts on children's subjective well-being, with children reporting higher levels of shame or embarrassment because of being shorter than their peers, and so is likely to impact on children's progression through school (Dercon 2008). It can also increase susceptibility to physical illness, as the body is less able to fight infections. Stunting therefore has long-term implications for children and also for the wider society, through lost potential.

As with inequalities relating to education, certain groups of children are more likely to be stunted than others. In all countries except India (Andhra Pradesh), higher household wealth is associated with greater chance of having a healthy height-for-age.[6] Similarly, in all four study countries children in urban areas are more likely to have a healthy height-for-age than their rural counterparts, with children in rural areas having lower access to quality services. Ethnicity is also a factor: there is a higher prevalence of stunting in Peru, Vietnam and Andhra Pradesh among children from ethnic minorities or lower-caste status, when controlling for other factors. Children who belong to the Scheduled Caste in Andhra Pradesh are 70 per cent less likely to have a healthy height-for-age, whereas in Vietnam children are 3.3 times more likely to be in the healthy range if they belong to the Kinh (majority) ethnic group. While groups with low social status tend to live in rural areas, the stunting rates for these groups are higher than the average for rural areas. This may be due to the difficulties associated with providing quality services in remote areas. It may also be a result of political economy whereby marginalised groups have less voice in demanding resources.

Poverty is also associated with lower subjective well-being. To collect data on children's perceptions of well-being, Young Lives asked them to position themselves on a ladder where the ninth step represents the best possible life and the first step represents the worst. Across all four countries, controlling for other factors, higher household consumption levels are associated with children positioning themselves higher on the ladder. This suggests that poverty and inequalities impact on children's subjective well-being. Notably, higher levels of maternal education are associated with a higher position on the ladder in all countries except Vietnam, and with higher self-rated health in all except Ethiopia.

The overlap between better outcomes in enrolment and formal learning and indicators of health is illustrated more clearly by Table 1. This disaggregates the number of cases of children in Peru with poor outcomes in one of the education and formal learning domains (enrol-

Table 1: Distribution of children poor in education and health, Peru, 2009

	Children not enrolled or poor cognitive achievement AND out of healthy range of height-for-age or BMI-for-age (n. = 308)	
Group	**%**	**Number of cases**
Boys 53.6% of sample	52.6	162
Girls 46.4% of sample	47.4	146
Rural 23.5% of sample	59.4	183
Urban 76.5% of sample	40.6	125
Mestizo 78.8% of sample	70.6	214
Ethnic minorities 17.3% of sample	25.4	77
White/Asian 4% of sample	4	12
Consumption quintile 1 (lowest)	26.4	81
Consumption quintile 2	21.2	65
Consumption quintile 3	20.8	64
Consumption quintile 4	17.9	55
Consumption quintile 5 (highest)	13.7	42

ment or low cognitive achievement) *and* in proxies for health (out of the healthy range for BMI-for-age or height-for-age). The key issue is whether children from particular groups are over-represented.

Stark inequalities are highlighted by the over-representation of children from rural areas, ethnic minorities and households in the poorest quintile of household consumption. The make-up of the group of children showing less good enrolment, cognition, and health indicators compared with the whole sample suggests that children from rural areas are 2.5 times less likely to fare well in life chances in both education and health. Similarly, children from ethnic minorities (1.5 times) and children from households in the poorest quintile of household consumption (1.3 times) are less likely to fare well. A similar picture emerges in Vietnam with children from ethnic minorities (over 1.5 times) and from households in the poorest quintile (1.3 times) being less likely to fare well in education and health, suggesting that in both countries there are highly marginalised groups.

In terms of scale, children with poor outcomes in one health-related and one education-related dimension are much greater in number in Andhra Pradesh and Ethiopia, with just over two-thirds of children in the older cohort in Andhra Pradesh and just under two-thirds of children in Ethiopia falling into this category. However, children from the poorest quintile of household consumption are only slightly over-represented in Andhra Pradesh and Ethiopia and not to the same extent as in Peru and Vietnam. Therefore, although poverty rates have declined in all four countries (Dornan 2010) this has not necessarily resulted in the reduction in cumulative and compounded risk for the poorest and most marginalised groups in society. It demonstrates also the way in which risk is mediated through poverty and structural disadvantage.

Poverty and inequalities are at the root of risks for children

Children and families living in chronic poverty have fewer resources to cope with risks or adverse events such as illness and death, which therefore hit them harder and can perpetuate the cycle of poverty and risk. At the same time, where possible households draw on a range of strategies to cope with and respond to risk. Children play a key role in managing risk within the household. The following section explores the interconnections between risk, protective processes and children's life chances, through the lens of family illness and death, selected because of their prevalence in the everyday lives of poor children and families.

Illness is both a cause and consequence of poverty, due to the poor-quality living environment of poor children and the indirect (lost income and time) and direct costs (user fees at healthcare facilities, cost of medicines and transport to the facility) of illness. Illness also tends to co-occur with other types of adverse events, such as food price increases, crop failure, and drought. For example, in Ethiopia 52.6 per cent of households reporting an environmental shock also reported experiencing a family illness or death, and 53.9 per cent of households reporting an economic shock in 2009 also reported experiencing a family illness or death (Dornan 2010: 9). The co-occurrence of adverse events and the impact of direct and indirect costs are illustrated by the situation faced by Harika's family.

Harika lives in a rural community in Andhra Pradesh with her parents and two brothers. Her family belongs to a Backward Caste. Harika's eldest brother was unwell and taken to a doctor in the town and initially diagnosed with malaria. When his condition did not improve, he was admitted to a private hospital in Hyderabad. After a series of tests, it was discovered that there was '*no circulation of blood on one of the nerves on [his] neck*'. The entire ordeal lasted for one month, during which time Harika's mother stayed by her son's side and was unable to work the family's fields. According to Harika: '*We did not take care of the cotton fields and that was the reason the crop was not good*'. The total cost of the treatment was 150,000 rupees, which the family paid by obtaining a loan at a low interest rate. Harika's maternal cousin helped to transport the family to Hyderabad and provided accommodation during their stay. On another occasion, Harika's father was unwell and Harika had to miss school in order to help her mother tend the fields. Harika feels that her household is poorer as a result of these events.

Family illness and death can have long-term impacts on children. For example, controlling for other factors, children in Andhra Pradesh who lost a family member (other than a caregiver) are 70 per cent less likely to be enrolled in school. Evidence from the qualitative data suggests that this is due to children having to take over the roles and responsibilities previously undertaken by the family member. In both Ethiopia and Vietnam the death of a father is associated with lower cognitive achievement, perhaps suggesting periods of absence or permanent drop-out from school. Therefore, ill-health and illness need to be examined at the household level. This builds up a more complete picture, as not only does illness of one member impact on all the others, the entire household including children is involved in managing ill-health, as illustrated by Harika's family.

In managing ill-health and illness, children and families depend on multiple sources of support. Children and caregivers described employing a range of strategies. These may include reducing consumption, selling assets, or using savings; labour substitution (by children or others), borrowing; seeking assistance from relatives, neighbours or NGOs; and/or use of social protection schemes. Families often employ several strategies simultaneously, as illustrated by the case of Denbel.

Denbel lives in urban Ethiopia with his single mother and younger brother. Denbel had malaria and he was taken to a private clinic, tested and provided with medication. The total cost of the treatment was 30 birr, which Denbel's mother '*borrowed ... with interest from our association in which we save money for buying meat for holidays*'. The daily amount of interest on 30 birr is 90 cents and Denbel's mother repaid half of the loan within a week, reducing the daily interest to 45 cents. A neighbour provided Denbel with support during his recovery:

> *The doctor told him to take the medication with milk. And [the] neighbour, who owns a cow, gave him one bottle of milk every day.*

Denbel's case illustrates also the role of semi-informal community-based organisations (such as the self-help groups in Andhra Pradesh or burial societies called *idirs* in Ethiopia). Controlling for other factors, household membership of these groups is associated with better outcomes for children, including enrolment (Andhra Pradesh and Ethiopia), health (Andhra Pradesh, Ethiopia and Peru), and subjective well-being (Ethiopia). Access to informal networks enables families to cope better with illness.

However, there are also dangers associated with debt-traps, and ability to access credit is often dependent on the household's social standing within the community. As one caregiver in Ethiopia explained, '*Only those who are trusted by the kebele leaders can receive credit*'.[7]

Consequently access to these protective processes is frequently determined by the same inequalities of power and resources (such as income, livelihoods, assets, credit, insurance, quality of the living environment, and access to quality services) that give rise to higher risk for poor families. This suggests a potential role for social protection, working alongside, rather than replacing, existing structures.

Rajesh is from a Scheduled Tribe and lives in rural Andhra Pradesh. His father contracted malaria and typhoid and was unable to work for a month. Faced with destitution, the family adopted a number of strategies. Rajesh and his siblings undertook work as part of the National Rural Employment Guarantee Scheme.[8] Rajesh did not have his own work card, so used his sister's during the holidays. Rajesh's elder sister also sought to help the family by providing 500 rupees for medical costs and his mother borrowed 1,500 rupees, with one per cent monthly interest, from a self-help group. As illustrated here social protection schemes provide poor families with more options, acting as an insurance or redistribution system, rather than a credit system, which entails debt.

Conclusions

While child protection is concerned with reducing risk and improving protection for children, it has the tendency to take the approach of enabling individual children to 'beat the odds' rather than 'changing the odds', which requires targeting the causes of poor life chances rather than just the symptoms (Seccombe 2002). Evidence from Young Lives challenges this approach by demonstrating that risk is driven by poverty and structural inequalities, repeatedly putting at a disadvantage the same groups of children, who fare less well across a series of indicators in education, health, and well-being. To protect children therefore, child protection needs to look beyond violence at the interpersonal level, to violence at the structural level.

However, there is the danger that child protection could be tasked with everything and consequently achieve nothing. An alternative and more feasible approach would be to place children at the centre of development debates and policies, integrating child protection concerns around sources of risk and protection. This would enable a shift from reaction to prevention and the injection of a political-economy perspective to understand how broader structural inequalities put children at risk. It would facilitate greater linking of the situation of children with that of their families and communities: risk and protective factors for children cannot be understood in isolation from families and communities. As poverty is compounded by multiple and recurrent adverse events, such as illness, there is also a need for policies that do not target one risk alone, since poor outcomes in one area, such as health, can undermine progress in another, such as education. Children and their caregivers offer insights into how formal schemes and informal networks do and do not work in mitigating the impact of poverty and adverse events on the household and should be used to inform policy. It is only by 'changing the odds' through such an approach that child protection will truly live up to its name.

Acknowledgements

My thanks are due to the children and families who participate in the Young Lives study and to the data-gathering teams in the study countries. I am especially grateful to Laura Valadez and Daniel Vujcich for assistance with data analysis and to Paul Dornan for support in developing the article.

Notes

1. Save the Children's definition of child protection, available at http://sca.savethechildren.se/PageFiles/3189/child%20protection%20definition%20SC.pdf (retrieved 29 September 2011).
2. Details can be found in technical notes which describe the sample, available at www.younglives.org.uk/our-publications/technical-notes (retrieved 29 September 2011).
3. The term 'caregiver' has been used in recognition of the fact that children are not always cared for by their biological parents.
4. Receptive vocabulary (school readiness), verbal skills, and achievement were measured using tests Young Lives developed from standardised international tests, such as the Peabody Picture Vocabulary Test (PPVT). Measures were adapted to local contexts and languages to minimise bias, an especially important consideration in testing children who speak minority languages. Reliability and validity results for test administrations and concerns are presented and discussed in Cueto, Leon, *et al.* (2009). A composite cognitive achievement score was created by coding separately PPVT and maths scores into low, medium, and high scores and then combining these variables.
5. The picture for children from ethnic minorities could be even bleaker. Regression analysis showed that if wealth were the same for both majority and minority groups, the majority Kinh group would be 80 per cent less likely than children from ethnic minorities to be enrolled in school.
6. BMI-for-age and height-for-age were calculated using international norms established by the World Health Organization. Children classified as having a healthy BMI and height fall within two standard deviations of the norm.
7. Kebele is the smallest administrative unit in Ethiopia.
8. The National Rural Employment Guarantee Scheme (NREGS) guarantees hundred days per year of wage employment doing unskilled manual work for adults in rural households. Although it is illegal for people under 18 years old to participate, children sometimes use family members' job cards to work on the scheme.

References

Boyden, J. and E. Cooper (2009) 'Questioning the power of resilience: are children up to the task of disrupting the transmission of poverty?', in J. Addison, D. Hulme, and R. Kanbur (eds.) *Poverty Dynamics:*

Measurement and Understanding from an Interdisciplinary Perspective, Oxford, UK: Oxford University Press.

Crivello, G. and J. Boyden (2011) 'Situating Risk in Young People's Social and Moral Relationships: Young Lives Research in Peru', Working Paper 66, Oxford, UK: Young Lives.

Cueto, S., J. Leon, G. Guerrero, and I. Muñoz (2009) 'Psychometric Characteristics of Cognitive Development and Achievement Instruments in Round 2 of Young Lives', Technical Note 15, Oxford, UK: Young Lives.

Cueto, S., G. Guerrero, J. León, E. Seguin and I. Muñoz (2009) 'Explaining and Overcoming Marginalisation in Education: A Focus on Ethnic/Language Minorities in Peru', background paper for *EFA Global Monitoring Report 2010*, Paris, France: UNESCO.

Dercon, S. (2008) 'Children and the Food Price Crisis', Policy Brief 5, Oxford, UK: Young Lives.

Dornan, P. (2010) 'Understanding the Impact of Crisis on Children in Developing Countries: Round 3 Preliminarily Findings', Oxford, UK: Young Lives.

Kabeer, N. (2000) 'Inter-generational contracts, demographic transitions and the "quantity-quality" trade-off: parents, children and investing in the future', *Journal of International Development* 12: 463–82.

Le Thuc, D. (2009) 'The Effect of Early-age Stunting on Cognitive Achievement among Children in Vietnam', Working Paper 45, Oxford, UK: Young Lives.

Levine, R., C. Lloyd, M. Greene, and C. Grown (2008) *Girls Count: A Global Investment and Action Agenda*, Washington, USA: Center for Global Development.

Luthar, S., D. Cicchetti, and B. Becker (2000) 'The construct of resilience: a critical evaluation and guidelines for future work', *Child Development* 71: 543–62.

Pells, K. (2011) 'Poverty and Gender Inequalities: Evidence from Young Lives', Policy Paper 3, Oxford, UK: Young Lives.

Sanchez, A. (2009) 'Early Nutrition and Cognitive Achievement in Pre-School Children in Peru', Working Paper 57, Oxford, UK: Young Lives.

Seccombe, K. (2002) '"Beating the odds" versus "changing the odds": poverty, resilience, and family policy', *Journal of Marriage and Family* 64: 384–94.

UNESCO (2010) *Reaching the Marginalised: EFA Global Monitoring Report 2010*, Paris, France: UNESCO.

UNICEF (2010) *State of the World's Children 2011: Adolescence – An Age of Opportunity*, New York, USA: UNICEF.

Werner, E. and R. Smith (2001) *Journeys from Childhood to Midlife: Risk, Resilience and Recovery*, Ithaca, USA: Cornell University Press

Action research exploring information communication technologies (ICT) and child protection in Thailand

Philip H. Cook, Cheryl Heykoop, Athapol Anuntavoraskul, and Jutarat Vibulphol

Traditional approaches to protecting children are insufficient to meet the complex issues they now face, and inter-sectoral, child-centred strategies are needed. Addressing this, the International Institute for Child Rights and Development (IICRD) developed the Circle of Rights (COR), a participatory action research approach to involve children in 'bottom-up' protection work. This paper describes COR in Thailand through the Child Protection Partnership (CPP), a project focused on ICT child protection. Children, youth and adults in four Thai communities collaborated in co-creating integrated strategies to address ICT and child protection. The paper describes the process and actions resulting from this child-centred partnership.

Recherche-action portant sur les technologies d'information et de communication pour l'information (TIC) et la protection de l'enfance en Thaïlande

*Les approches traditionnelles de la protection de l'enfance ne sont pas suffisantes pour répondre aux questions désormais en présence et ce sont des stratégies intersectorielles et centrées sur l'enfant qui sont nécessaires. Pour remédier à ce problème, l'Institut international pour les droits et de développement de l'enfant (*International Institute for Child Rights and Development - *IICRD) a mis au point le Cercle des droits (*Circle of Rights - *COR), une approche participative basée sur la recherche-action afin de faire participer les enfants au travail de protection « ascendant ». Cet article décrit le COR en Thaïlande à travers le Partenariat pour la protection de l'enfance (*Child Protection Partnership - *CPP), un projet portant sur la protection de l'enfance dans le contexte des TIC. Les enfants, les jeunes et les adultes de quatre communautés thaïlandaises ont collaboré pour créer ensemble des stratégies intégrées afin d'aborder les questions liées aux TIC et à la protection de l'enfance. Cet article décrit le processus et les actions découlant de ce partenariat centré sur l'enfant.*

Pesquisa de ação explorando tecnologias de comunicação da informação (ICT) e proteção infantil na Tailândia

Pesquisas tradicionais de proteção às crianças são insuficientes para tratar das questões complexas por elas enfrentadas atualmente, sendo necessárias estratégias intersetoriais, centradas nas crianças. Abordando isto, o Instituto Internacional para Direitos da Criança e Desenvolvimento (IICRD) desenvolveu o Círculo de Direitos (COR), uma abordagem de pesquisa de ação participativa para envolver crianças no trabalho de proteção de "baixo para cima". Este artigo descreve o COR na Tailândia por meio da Parceria de Proteção Infantil (CPP), um projeto concentrado na proteção infantil da ICT. Crianças, jovens e adultos em quatro comunidades tailandesas colaboraram na criação conjunta de estratégias integradas para abordar a ICT e a proteção infantil. O artigo descreve o processo e as ações resultantes desta parceria centrada na criança.

La investigación-acción sobre las tecnologías de información y comunicación (TIC) y la protección de la niñez en Tailandia

Los métodos tradicionales que se utilizan para proteger a los niños y niñas son insuficientes para enfrentar los complejos problemas de la actualidad; se requieren estrategias intersectoriales y centradas en la niñez. Para enfrentar esta problemática, el Instituto Internacional para los Derechos del Niño y el Desarrollo creó el Círculo de Derechos (CdD), un método participativo de investigación-acción en el que los niños y niñas participan "desde la base" en acciones encaminadas a su protección. Este ensayo describe un CdD que creó la Alianza para la Protección de la Niñez en Tailandia, un proyecto que se centra en la protección de la niñez a través de las TIC. Los niños, niñas, jóvenes y adultos de cuatro comunidades tailandesas se unieron para diseñar conjuntamente varias estrategias que integran la protección de la niñez y las TIC. Este ensayo describe el proceso y los resultados de esta alianza centrada en la niñez.

Introduction

A number of global changes, including the growth of information and communications technologies (ICT), and shifting demographics (aging populations in the Global North and youthful demographics in the Global South) implicate children and childhood in multiple and important ways (Lynch 2010). Responding to these global changes presents both challenges and opportunities to child protection – a relatively youthful social service sector compared with say health or education. The challenge of devising effective protection services is especially acute where new kinds of problems for children and youth arise as a result of evolving ICT. The following paper describes a participatory research journey exploring ICT-enabled child protection in Thailand. Multiple perspectives from children and youth and other adult stakeholders in four Thai regions weave different themes as a complex narrative about the impact of ICT on children and the ways various social sectors can respond to threats arising from ICT. The paper also charts the emerging opportunities for children and youth to be agents of change in their own protection and in the creation of new models of child protection through a participatory process of critical reflection and engagement.

The Asian region has the most rapidly growing population of ICT users in the world, most of whom are young people between the ages of 14 and 25. In the last ten years, Thailand has seen staggering ICT user growth of 660 per cent, with the rate rising from two to 17 million people. As a youthful country, this has important implications for the ways children and youth will shape the future of Thai social, economic and political systems (Internetworldstats.com 2011).

ICT is an umbrella term including any communication devices or applications: radio, television, mobile phones, computer, hardware and software, satellite systems, and services and applications related to them such as online chatting, and social networking (IICRD 2011). These technologies are constantly evolving, merging, and creating new hybrids and applications; for example mobile phones, Internet, video conferencing, and social networking are all being applied together in the evolution of smart phones.

According to ECPAT (2008), an international network of organisations and individuals that brings together research and advocacy to eliminate the commercial sexual exploitation of children, one of the fastest growth areas in ICT use is online pornography, including the phenomena of online paedophilia and the sexual abuse of children. Another current agent of great change and a fast growth area of ICT use is social networking amongst young people. Young people are therefore not simply victims in the ICT world. Many new ICT technologies are being developed by adolescents and youth, and young people are themselves transforming the ways in which ICT is defining the social, commercial, and political facets of human connectivity. This ranges from the mundane world of collective shopping 'groupons', through the creation of unique online identities and personal webs of relationships, to well-orchestrated populist social network movements aiming to overthrow non-democratic governments in the Middle East and other regions of the world.

The Child Protection Partnership (CPP)

To address these issues and develop innovative systems of social support, the International Institute for Child Rights and Development (IICRD) created the Child Protection Partnership (CPP). This global research initiative attempts to understand the unique and emerging threats posed by ICT sexual exploitation; it also examines how young people are involved in finding solutions to these risks and in applying the results of this research to strengthen social protection systems.

The goal of CPP is to reduce, and where possible eliminate, ICT-enabled child sexual exploitation by building on good practices and technology across public and private sectors. CPP has three entry points to address the issue:

- Equip law enforcement, government, and other supporting bodies or organisations better to address ICT-enabled child sexual exploitation.
- Connect vulnerable children (male and female) to protective mechanisms and services that prevent and address ICT-enabled child sexual exploitation.
- Foster a coordinated systems approach amongst stakeholders, supported by CPP to prevent and address ICT-enabled child sexual exploitation.

For CPP to succeed in strengthening child protection systems, interventions at all these entry points depend on adequate consultation and cooperation with the young persons involved. CPP encourages children to protect themselves and to contribute to peer prevention and protection through working with bodies such as governments, policymakers, police, owners of Internet cafés, and non-governmental organisations. To do this, CPP partners work with young people to build on their strengths and expertise in ICT child protection.

To understand the complexity of ICT and child protection, and develop contextualised solutions in local Thai contexts, the CPP team believed that direct engagement with children was necessary. This type of engagement required providing children with opportunities to reflect upon the problems facing them and develop viable solutions, and would also need to connect children to a broad range of stakeholders to help address children's concerns. Accordingly, a child-centred methodology developed by IICRD called 'The Circle of Rights' was applied, in which the foundations of children's human rights and community engagement with children are integral.

The Circle of Rights (COR): intentional planning in social change

Adopted in 1989, the United Nations Convention on the Rights of the Child (UNCRC) represents the most widely ratified human rights treaty in history, and has been adopted by every country except the United States of America and Somalia. The UNCRC articulates the minimum legal and ethical commitments and standards for optimal child development, and provides a framework for accountability and collective action to advance the human rights of all children. Despite widespread legislative and policy commitments, research suggests actions to implement the UNCRC still fail to realise the rights of many children in their day-to-day lives. Critics argue that these shortcomings result from the application of generic, imposed, or Western-centric approaches, affording limited regard to local social and cultural contexts and conceptualisations of children and childhood. Rather such critics call for innovative and creative inter-sectoral approaches to contextualise the principles of the UNCRC in ways that synchronise with the unique and changing realities of children, families, and communities, and demonstrate positive empirical change in the context of children's lives (Bissell *et al.* 2008).

To help ground children's rights locally and within the context of children's day-to-day lives, IICRD, a hybrid academic/non-governmental organisation located at the University of Victoria, Canada, has worked over the last 16 years to develop the Circle of Rights process (COR). This child-centred, participatory, action-oriented research process begins with children's perspectives and helps communities to build from local wisdom and community strengths to develop innovative and contextualised strategies to support positive social change for children reflecting the principles of the UNCRC.

The metaphor of a young fern growing is used to describe the COR research process and also represent the strength-based, iterative learning-action process embedded within the four distinct phases of COR: Map; Engage; Plan; and Create.

The COR methodology has been adapted and applied around the globe to tackle a variety of complex challenges in relation to child rights and protection, including HIV/AIDS, natural disasters, extreme poverty, the rights of indigenous children and their communities, violence and conflict, and the development of contextualised child protection and governance systems. While the principles of COR have remained constant, the practice of COR has reflexively evolved to integrate lessons learnt through application. In this paper we demonstrate the COR methodology by looking at its application in Thailand through CPP.

Methodology

Community partners

To better understand the issue of ICT-enabled child sexual exploitation in Thailand, CPP worked through local partners in three initial pilot communities:

1. Human Development Foundation and Mercy Centre in Klong Toey, a slum area in Bangkok
2. Plan Thailand International in Baan Nato, a community bordering Burma in the Province of Chiang Rai
3. We Peace in Grongpinung, a community affected by the conflict in the Yala Province in Thailand's Deep South

In the later stages, CPP also engaged with children in Pattaya, a community near Bangkok and one of Thailand's most notorious centres of commercial sex tourism. At a national level, there was collaboration with three key partners: the Faculty of Education of Chulalongkorn University, Plan Thailand International, and Fight Against Child Exploitation (FACE).

Participant selection

Participant selection was purposive, including factors of age, gender, populations at risk, and accessibility to ICT. While efforts were made to ensure diverse representation across these factors, sampling was also partially based for convenience on pre-existing relationships with local partners, and willingness to participate in the research. In total, 108 children and young people participated, including 27 children from Klong Toey (19 boys, 18 girls, aged 11–16 years); 51 children from Baan Nato (16 boys, 35 girls, aged 13–19); 30 children from Grongpinung (17 boys, 13 girls, aged 10–12). Later, work in these three communities was supplemented with focused activities in Pattaya. Adults from these communities also participated, including parents, officials in local government and law enforcement, and members of broader civil society.

The findings do not represent the views of all young people living in Thailand or even an accurate cross section. Rather, the intention is to bring together the voices and experiences of a sample of children, their communities, and key child protection partners (government, law enforcement, NGOs) to shed light on their experiences and perspectives of interaction in their community between child protection and ICT. A secondary purpose is to use this information to inform inter-sectoral and collective action to strengthen accountability and foster new approaches to child protection. The children participating in this process were generally local young people aged 12–18 who were already engaged in ongoing community activities organised by the host organisations.

Map phase

The first step in COR involved locating the issue of ICT-enabled child protection within the national Thai context of child rights. This includes mapping the child rights mandate (legislation, regulation and policies), inter-sectoral actors (duty-bearers and systems related to child rights and protection), and mechanisms (strategies, interventions, good practices and gaps) related to ICT and child protection. A variety of activities were used including: a desk review to identify relevant laws, policies and programmes addressing ICT and child protection, and key informant interviews to identify national mandates, inter-sector stakeholders and collaborations, and mechanisms (projects and innovations) related to ICT and child protection.

For this study the national child rights mandate included a review of Thailand's Constitution, the Child Protection Act, the Anti-Human Trafficking in Persons Act, the Penal Code, the Computer Crime Act, and the Decentralization Act. While some of these legal instruments provide broad protection such as the right of children to be protected from violence and unfair treatment, there is virtually no reference to ICT or related issues such as child pornography. However, some strengths were identified that may be built upon, such as provision to suppress dissemination of pornographic material (Computer Crime Act 2007).

The main finding of the Map phase was that, while Thailand has strong child protection laws and systems, the laws give limited focus to ICT and the corresponding implications on child protection. In addition, the desk review highlighted that the inter-sectoral committees established under the Child Protection Act provide a unique opportunity to inform policy development and adopt innovative inter-sectoral strategies to address ICT-enabled child protection and prevention of child abuse. However, there remained a lack of clarity with regards to the agencies responsible for establishing and enforcing procedures and operations in game shops and internet cafés, and this lack of clarity may present a challenge.

In regards to the policing policies relevant to ICT-enabled child protection, the mapping revealed that Thailand's newly established Crime Suppression Division (2009) is responsible for implementing the Computer Crime Act. Unfortunately, this division of the Royal Thai Police (RTP) has received limited training and capacity building on child protection and its relationship to ICT beyond paedophilia.

Finally, the mapping highlighted an opportunity to strengthen the role the wider community plays in addressing ICT-enabled child protection. Under the Decentralization Act, the Department of Local Administration (DLA) holds responsibility for the creation of community action plans for child and youth, and is also responsible for local policing, education and youth engagement through the establishment of a local youth council (Varanyuwatana 2009). This presents a unique chance to work with the DLA to develop innovative, inter-sectoral, youth-led solutions at the community level.

Engage phase

With the Map phase complete, CPP partners began the Engage phase of COR through engaging with children, their communities, and child protection system stakeholders to understand perspectives on ICT and child protection, and to identify the associated challenges and opportunities. Specifically, the Engage phase sought to understand the perspectives of children and other stakeholders including families, community leaders, law enforcement officers, governments, and child protection organisations.

Activities started with collective, playful, trust building, and individual exploration activities; this was a first step to create more open and critically reflective dialogue with children. These activities later served as a platform for understanding the issues from the various perspectives.

Activities were organised in a layered, progressive manner, with each activity building from the previous one, with an increasing depth of inquiry on critical ICT risk and protection issues. These sessions began by creating a safe place for explaining the purpose of the research with children and for young people to share their personal experience with facilitators; specifically, the sessions allowed children to explore their inner worlds through their thoughts, feelings, and perceptions about themselves and their interaction with others. Consideration was then given to children's outer worlds. These activities explored young people's understandings of the world around them and how they considered community (family, friends, people in community, culture and traditions) and institutional relationships (with schools, organisations, and government) within their social and physical environments (including places they go to, people they see, activities they do). Next was an exploration of children's individual and collective interactions with diverse aspects of ICT, including mobile phones, computers, video, and the Internet. In all activities, an emphasis is placed on safety, imagination, critical inquiry, and mindful, insightful, action-oriented reflection and discussion.

Understanding children's lived realities of ICT

This section outlines key themes that emerged from the Engage phase with children and youth from Klong Toey, Baan Nato, and Grongpinung, and youth in Pattaya.

Limited capacity of adults to provide ICT advice and safeguards

While the children identified the importance of receiving advice from adults in their lives, it became evident through the conversations that this is challenging in the context of ICT, where children, not adults, are often the experts. Children reported that many of their parents or guardians do not understand how to use ICT and do not monitor children's usage; it is therefore difficult for parents and adults in the community to provide proper guidance on safe ICT use. Schools are more aware of ICT benefits and risks, yet the implementation of safeguards to protect children remains limited. For example:

At school the computer room is open from 8 am to 8 pm. But we are not allowed to chat, to watch clips, to play games, to watch movie stars' pictures. Everyone has a password to use computers. If someone sees things that they are not supposed to, teachers will know. (14-year-old boy, Baan Nato)

Yet during focus-group discussions, these children and youth also indicated that the school does not have a programme to block inappropriate sites and that children and youth can find (either intentionally or accidentally) inappropriate clips, games, and images.

In some situations, safeguards exist to help keep children safe in Internet cafés and game shops. In Pattaya, regulations prohibit children from entering such establishments when wearing their school uniforms; the same regulations also identify hours when children are permitted to use the cafés. Yet, according to children and community police, these regulations are rarely enforced and also easy to avoid. For example, one child described how he changes his shirt before he enters the café. Children and young people also explained that the café and shop owners want their business and may turn a blind eye should someone under-age enter the shop. One disturbing Internet café trend mentioned in the Pattaya case study was the phenomenon of school-aged girls being encouraged to wear their school uniforms when using the cafés, in order to entice older foreign tourists with a sexual appetite for school-aged girls to use the cafés at the same time.

Mobile phones

Through the ICT mapping activity, children explained that mobile phones can be used in many ways, including:

- staying connected to friends and family
- taking pictures and recording video clips
- playing music and games
- having access to GPS locators or an alarm clock
- most important, to connect to the Internet.

Mobile phones were also being used to record and distribute graphic sexual or violent images, which can be used innocently or with intention to harm other children. Conversely, mobile phones were frequently used as a protective mechanism to call for help or to warn others of danger.

Mobile phones are the most accessible form of ICT for children. All children in Klong Toey have access to mobile phones, with 17 of the 28 COR participants owning their own. Even in remote Baan Nato, 20 of the 45 participating children either own or have access to mobile phones. Children and youth primarily use them to communicate with family and friends and to discuss any problems they may be facing.

I use it [mobile phone] to receive calls from my sisters and my neighbours; I use it to call my Mom and sister; I talk with friends on the weekends (13-year-old girl, Baan Nato)

Furthermore, mobile phones help foster a sense of belonging and identity amongst peer groups, reflecting a child's social status:

I want someone to ask for my mobile phone number so other people will see me as important. (14-year-old girl Klong Toey)

A child living in a poor community may have difficulty in accessing or using a mobile phone. Children in Klong Toey and Pattaya spoke of the various ways they can obtain mobile phones, including through drug dealing, stealing, and exchanging sexual favours. One community facilitator described how one young girl exchanged sex for money to buy a mobile phone, and how another young man steals money to buy phone credit and play online games. While this information is anecdotal, the information shared in confidence with the trusted key informant reveals a theme that emerged in other key informant interviews, especially in the larger urban centres of Bangkok and Pattaya. It is clear that access to ICT comes at a significant personal and social cost for some, and issues of ICT and child protection cannot be considered as separate from the broader issues of poverty.

In the context of violence in Thailand's southern insurgency, mobile phones can also be a particularly important protective tool for a child. For instance, one young woman described an incident in which a young boy was in his room and heard gun shots and the screams of his parents as they were shot. With his mobile phone he was able to call the police. The young woman suggested:

> *Had it not of been for the mobile phone call the boy would have been killed. (17-year-old girl, Grongpinung)*

Another young woman, who had lost all her siblings in the conflict, reflected on how important ICT has been in her life:

> *I survive because of my mobile phone. My family is all dead. When anything happens, I can survive because of my mobile phone. I can call the soldiers. (18-year-old girl, Grongpinung)*

Yet mobile phones can also pose considerable threats for children in their communities. The young women we spoke to in Grongpinung described how mobile phones are sometimes used to activate bombs by youth extremists. They also described how GPS applications can be used by authorities to track people.

ICT and adolescent sexuality

One of the ways children and youth explore sexuality and personal relationships is to engage in acts of 'sexting' (sexualised texting). This involves sending sexually explicit text messages or photographs to ICT users, primarily through mobile phones.

According to the participating children and young people, sexting is commonplace. In their opinion, sexting can boost self-esteem and it can feel good to flirt with others. For instance, sending an image of oneself naked can illicit praise or compliments, or can simply show that you care for someone. Young people, however, also talked about the negative implications of sexting and discussed what can happen when these sexual images get into the wrong hands. For instance, young people from Klong Toey explained that if a girl breaks up with her boyfriend, he can send the image to others. Or other people can record images and video clips and disseminate them widely without the permission of those involved, leading to deep personal shame and humiliation.

Mobile phones are now equipped with an array of tools to capture video clips and images, and children and young people use these tools widely to document their lives and share with others. Unfortunately, these images can also violate privacy in ways that may be detrimental to the lives of children. We heard many stories from young people to this effect, where sexual images were recorded and widely distributed. Various perspectives on this issue included:

> *I once received a video of a student being raped and the clip was transferred to other friends. (15-year-old boy, Klong Toey)*
>
> *There was girl in sixth grade having sex with a secondary school student in the park. It was*

recorded and was sent to other students. When the school found out the girl was expelled. They said that she was giving the school a bad reputation. I knew that girl. (15-year-old girl, Klong Toey)

A ten-year-old girl had sex with a twelve-year-old boy. A friend recorded it and sent it to his friends. It was spread throughout the village, and the girl's dad saw the recording. The father was angry but what could he do? When the child was asked about the event by the supporting organisation, she said it did not matter. She had been in similar circumstances before where sexual images of her had been circulated. She said, 'It is just is how it is.' (14-year-old girl, Grongpinung)

The importance of safe, creative and affordable online access

The importance of feeling safe in ICT spaces was a common theme, and children and youth reiterated their desire for a safe place to access the Internet. This value was shared generally by all children across the Thai sites and was captured in the following quote from Pattaya.

We want to be able to access Internet in a place where we feel safe. We want a place that is inexpensive and where we don't have to see pornographic images. (14-year-old girl, Pattaya)

There was some variation in the type of safety being sought by the young people participating in the study, with a greater value being placed on safety from community conflict and violence in the Deep South, safety from pornography being highlighted in Pattaya and Bangkok, while in the North safety was more generally defined in regards to a broader cross section of issues including violence, abuse, and exploitation.

Unfortunately, many of the children and youth we spoke with across Thailand are unable to access computers or the Internet at home or at school and rely on community game internet shops, where governing laws are often scarce or non-existent; many shops are not registered. Children in Klong Toey and Pattaya talked extensively about the challenges of accessing safe, creative, and affordable Internet and gaming shops, and a variety of other child protection concerns. They talked about how some game shops are used to trade drugs and how they are surrounded by adults accessing pornographic images or chats:

There is a man who sits at the back of the Internet café and chats through a webcam with nude women. (15-year-old boy, Klong Toey)

Children described how shop owners shared pornographic images with them and encouraged them to chat with strangers. They also talked about how owners of game shops use promotions or incentives to encourage or lure young people into particular shops (such as 'buy two hours and get one free', 'free food and drink if you play all day', or 'bring a friend and you can play for free').

The impact on children and of Internet pornography and violence

Young people participating in the Circle of Rights process eloquently described how ICT can serve as a tool to expose children and youth to social influences for which they may not be prepared. In this emerging online environment children may be senders, receivers, victims, or active perpetrators of ICT-related sexual and violent acts. Several other examples were raised by law enforcement personnel, and these are identified in the next section.

Children in Klong Toey talked about the widespread distribution of pornographic images and clips. This phenomenon is so widespread that they described how amongst younger children this is now becoming a normalised aspect of their lives. According to the facilitator, the children and youth frequently talked about pornographic clips:

They [porn clips] were the first thing children mentioned, and they [the children] described how they use mobile phones to watch porn clips and transfer these clips to friends. (Adult facilitator, Klong Toey)

Another young person described how, while she was recently online, someone had wanted to chat with her:

Last week I was using MSN. A guy said he was from Yala and wanted to talk to me. I said I didn't have a camera, but I accepted access to his camera. When I saw his video stream it was his penis on line. He was proposing to have sex. He was much younger than me. I threatened to call the Ministry of Culture. He left. (16-year-old girl, Grongpinung)

In conflict, ICT can also help normalise violence and violent behaviour. Young women from the Deep South described an instance where the beheading of a soldier was recorded on Bluetooth,[1] and the video was widely disseminated.

It [the recording] was so easy to find. It makes violence normal. (15-year-old girl, Grongpinung)

Children are often seen re-enacting violent events, such as killings, beheadings, and shootings in their games and other peer-related social activities. According to one young woman:

Children are taking this [violence] all in. Children are seeing soldiers using guns. It is influencing the way they behave, and it is even worse when children have no guidance from parents. (15-year-old girl, Grongpinung)

They often take pictures of soldiers carrying guns. It makes them feel safe and secure. (14-year-old girl, Grongpinung)

In conflict, ICT can save, replicate, and disseminate violent images and may further contribute to a further normalisation of violence in communities.

The complexities and opportunities of ICT for young people

It is clear that the interface between children, youth and technology is complex and goes beyond the boundaries of a child protection paradigm focusing solely on risk and protection. Much of the research took place against the backdrop of the Thai political struggle between the 'red' (generally in Northern Thai poor regions demanding greater democratisation) and 'yellow' (urban-based monarchist) movements, in which many young people were actively involved, using the Internet to express their social and political views.

Many of the focus-group collective conversations also addressed areas where young people felt a need to bridge the best of traditional Thai wisdom and cultural values (such as the concept of *Samakhi*, or collective unity) with the best of modern laws, protective agencies, and youth participation.

Plan and create phases

The Plan and Create phases of COR provided opportunities for children, communities, and implementing organisations to review the results of the Engage phase and begin establishing priorities, ideas, and strategies for community action related to ICT and child protection. The Plan phase identified key themes and preliminary strategies with children and adult stakeholders, and the Create phase provided a forum in which children and communities could provide input on tangible actions to strengthen child protection and other government and non-governmental systems.

A central aspect of the Plan and Create phases was the creation of negotiated spaces, or opportunities for critical reflection amongst children and adults, on themes emerging from the Engage phase of the Circle of Rights. A key aspect of the facilitation of these discussions is an examination of the roles of government and non-governmental agencies, duty bearers, family and community stakeholders, and children in creating a stronger child rights environment, part of which involves in-depth reflection on the issues raised by children and an identification of the responsible organisations and government departments.

Activities in the Plan and Create phases included a variety of participatory reflective planning tools. One example is the creation of Strength and Challenge charts where participants organise themes identified in the engage activities into 'strengths', 'challenges', or a 'grey zone'. The grey zone represents things that could be both a strength and a challenge (for example sexting can strengthen a positive sexual identity but can also harm that identity when information is shared without that person's permission). Another activity involved creating a 'Strategy Tree', where children were encouraged to identify and reflect on a key challenge, and identify the root causes of the challenge as well as some possible solutions or strategies. This activity also encouraged children to be reflexive and action-oriented in their thinking, for example, they considered how strengths outlined in the engage phase could be used to design solutions or strategies. Once action plans were identified with children and youth, community consultations were then hosted to explore the action plans developed, to establish inter-sectoral round tables, and to bring together children and their communities with local government and NGO representatives.

Child and youth priorities and proposed strategies

While the ICT landscape is quite different in each context, children in all four communities identified educational issues related to ICT and child protection as a key priority for the development of their action plans, reiterating a need to be better equipped and informed about safe use. In Baan Nato, children indicated that because ICT access is just becoming accessible, they have a particular need to better understand the benefits and risks of ICT so they can protect themselves when new threats emerge. In Pattaya and Klong Toey, where access to ICT is far more pervasive, children identified a need and desire to use the information gained during their involvement in CPP to share information on benefits and risks of ICT with others. This action idea was subsequently taken up by Plan International Thailand, IICRD, and the Thai Ministry of Education, creating and piloting a high-school curriculum on the safe and creative use of ICT.

Children in Pattaya also proposed ideas for action, specifically addressing the local threats of commercial sex tourism and other forms of abuse, and in the Deep South young people wanted better school-based education on self-protection from violence. Other actions emerging from the COR included: strengthening youth social networking with out-of-school young people for self-protection – which is being addressed by Plan International in partnership with the Mekong Youth Net, a youth-led network of NGOs in the Greater Mekong sub-region that is working on issues of self-protection from sex trafficking; and youth media coalitions in the Deep South. The final action plan that was created from the COR process was a partnership between youths and adults in Klong Toey and Pattaya, focusing on developing greater

Figure 1: The circle of rights (from IICRD 2010: 1)

cooperation between community police, children, and game-shop owners in order to create a 'Community Code of Conduct for the Safe Gaming and ICT Centres'. One unintended early consequence of this action is that Plan International has also worked with some of these community game centres to create community-based ICT hubs for broader family use, including special programmes for young mothers to access current information on pre- and post-natal care.

Finally, a national outcome of the CPP process was the creation of an inter-sectoral committee on child protection and ICT, chaired by the National Institute for Child and Family Development at Mahidol University. This committee currently reports to the national child protection committee and is providing input on legislative and policy reform.

Conclusion

The Thai CPP experience underscores the complexity of understanding and responding to emerging threats to children in ways that both meet the changing reality of children in the world and develop systems of support that are responsive to context. The COR process was clearly limited in scale and time span. Neither are the findings representative of Thailand's complexity as a country, nor can the emergent actions speak to the challenges of long-term sustainability of their impact on children and their protection. The COR process does, however, point to a promising direction in regards to bottom-up approaches to the implementation of the UNCRC in ways that involve children in not only defining problems, but also identifying local and collective assets that can be built upon to create more responsive solutions.

A summary of key areas where COR can assist in strengthening child protection systems includes:

1. understanding the existing child protection context and identifying opportunities for influencing legislative and policy frameworks
2. developing family, community or school-based activities that give visibility and voice to children and young people, and highlight their unique experiences with, and perspectives on, child protection
3. creating negotiated community social and conceptual spaces for shared understanding and critical reflection around identified child protection concerns
4. supporting informed collective action and social innovation

While the COR process can assist in strengthening child protection systems themselves, the participatory, inclusive, and action-oriented nature of COR can also impact directly on children, communities, organisations, and other partners involved. For example COR may assist young people, and other stakeholders, to recognise the value and importance of their unique perspectives, fostering more engaged citizens at the community level. Additionally, the information harvested through COR can assist organisations to improve their own organisational programmes and practices, and strive to adopt the principles embedded in the COR process in their day-to-day lives.

Finally, while the mapping information gathered in this research speaks to the many challenges and gaps in the current legislative and policy environment on issues like ICT and child protection, the data gathered in the COR Thai process and subsequent policy round tables provide a platform upon which broad-based social protection policy can be created with children playing a meaningful role.

Note

1. Bluetooth wireless technology is built into electronic gadgets and lets you talk and share information such as voice, music, and videos, wirelessly (www.bluetooth.com).

References

Bissell, S., J. Boyden, P. Cook, and W. Myers (2008) 'Rethinking Child Protection from a Rights Perspective: Some Observations for Discussion', unpublished report. CA: University of California Davis.

ECPAT (2008) 'Online child sexual abuse: The law enforcement response', Rio, Brazil: ECPAT International, available at www.ecpat.net/worldcongressIII/PDF/Publications/ICT_Law/Thematic_Paper_ICTLAW_ENG.pdf (retrieved 15 February 2012).

IICRD (2011) 'Exploring ICT Enabled Child Sexual Exploitation Through the Lived Realities of Young People', Child Protection Partnership/CPP, Victoria, Canada: IICRD, available at http://iicrd.org/system/files (retrieved/Circle_Thai_THAI%20FINAL%20June%2013.pdf (retrieved 26 April 2012).

IICRD (International Institute for Child Rights and Development) (2010) 'The Circle of Rights: Intentional Planning for Social Change', Victoria, Canada: IICRD, available at http://www.iicrd.org/research_circleofrights (retrieved 26 April 2012).

Internetworldstats (2011) available at www.Internetworldstats.com (retrived 25 April 2012).

Lynch, K. (2010) 'The global drivers of change', *Policy Options*, December 2009–January 2010: 75–77, available at www.irpp.org/po/archive/dec09/lynch.pdf (retrieved 15 February 2012).

Varanyuwatana, S. (2009) 'Decentralization and the Budget for Social Services at Tambon Administrative Level', Bangkok: UNICEF Thailand & Thammast University.

Child protection: a role for conditional cash transfer programmes?

Natalia Streuli

Drawing on empirical data from a community-based study with children and adults in rural Peru, the paper analyses the everyday experiences of a conditional cash transfer programme, called 'Juntos'. *The findings show that social protection programmes like* Juntos *address certain child vulnerabilities by making eligibility for their cash transfers conditional on behaviour-related to child protection-related such as health check-ups and school attendance. However, there are other aspects of children's well-being that are not being considered, such as experiences of violence and exclusion. This paper discusses both opportunities and challenges for cash transfer programmes to play a greater role in child protection.*

Protection de l'enfance : un rôle pour les programmes de transferts monétaires conditionnels ?
Grâce à des données empiriques tirées d'une étude communautaire avec des enfants et des adultes dans le Pérou rural, cet article analyse les expériences quotidiennes d'un programme de transferts monétaires conditionnels appelé Juntos. *Les conclusions montrent que les programmes de protection sociale comme* Juntos *remédient à certaines vulnérabilités des enfants en soumettant leur admissibilité aux transferts monétaires à des conditions relatives au comportement en matière de protection de l'enfance, comme les visites médicales et l'assiduité scolaire. Cependant, il y a d'autres aspects du bien-être des enfants qui ne sont pas pris en considération, comme les expériences de violence et l'exclusion. Cet article traite des occasions et des défis pour que les programmes de transferts monétaires jouent un rôle plus important dans la protection de l'enfance.*

Proteção infantil: um papel para os programas de transferência de dinheiro condicional?
Apoiando-se em dados empíricos de um estudo baseado na comunidade com crianças e adultos na zona rural do Peru, o artigo analisa as experiências cotidianas de um programa de transferência condicional de dinheiro chamado "Juntos". *Os resultados mostram que programas de proteção social como o* "Juntos" *abordam certas vulnerabilidades da criança ao tornar o direito às transferências de dinheiro condicional ao comportamento relativo à proteção infantil, tais como consulta médica e frequência escolar. Porém, existem outros aspectos do bem-estar das crianças que não estão sendo considerados, tais como as experiências de violência e exclusão. Este artigo discute as oportunidades e desafios dos programas de transferência de dinheiro para que tenham um maior papel na proteção infantil.*

¿Contribuyen los programas de transferencias condicionadas a la protección de la niñez?
Basándose en datos empíricos de un estudio comunitario con niños y adultos en el Perú rural, el documento analiza la práctica habitual de un programa de transferencias condicionadas denomi-

nado Juntos. *Las conclusiones muestran que los programas de protección social como* Juntos *abordan algunas vulnerabilidades de la niñez al considerar como criterios de elegibilidad para sus transferencias condicionadas temas relacionados con la protección de la niñez, como reconocimientos médicos y escolarización. Sin embargo, hay otros aspectos del bienestar de los niños que no se toman en cuenta, como la violencia y la exclusión. Este documento analiza las oportunidades y los retos para que los programas de transferencias puedan aportar más a la protección de la niñez.*

Introduction

Evidence from around the world shows that children are the age group most affected by poverty (UNICEF 2005). This is particularly the case in developing countries where over a billion children experience '*severe deprivation of basic human needs*' such as food, safe drinking water, health, education, and adequate shelter (Gordon *et al.* 2003: 25). In the light of this, a new generation of programmes specifically targeting children from poor households was introduced in Latin America during the 1990s. These programmes are social protection measures called 'conditional cash transfers' (CCTs), and rapidly became a critical component of poverty reduction strategies.

CCTs claim to address short-term income needs as well as to promote '*longer-term accumulation of human capital*' by providing money to families in poverty contingent upon certain verifiable actions and behaviours, such as regular school attendance or participation in basic preventative health care (de la Brière and Rawlings 2006). The intended benefits of CCTs for children include improvements in nutrition, education, health, birth registration and, in some cases, reduction of the number of children working (Bouillon and Tejerina 2006; Morley and Coady 2003; Valencia 2008).

Though promising, CCT programmes beg an important question about the interplay between objectives and benefits for children, both in the short and long term. By virtue of their 'human capital' rationale, CCTs tend toward a long-range focus on children as future adults, prioritising their preparation for adult life as economically productive citizens. The question is whether, in pursuit of children's future skills, these programmes may neglect children's experiences in the here and now. Although children's immediate health, educational progress, nutritional status and so forth may be tracked in the better programmes, some of which claim to be 'child oriented', this is often in terms of generalised 'indicators of progress', such as numbers of immunised children or children attending primary education.

However, there is another way in which to understand 'child centred', and that is as children's participation as partners in the design, implementation, or evaluation of CCT processes. Children are granted the human right by Article 12 of the UNCRC to participate in decisions about matters that affect them. Without such participation, little is likely to be known about how children and their families experience these programmes, and about the effects that they have on children's lives. The question is whether lack of inclusion of children's views and experiences may not only abridge children's rights, but also undermine the beneficial impact of CCT policies and activities.

This paper[1] explores how a CCT called '*Juntos*' ('Together') is experienced by children and their families in rural communities in Peru. The study draws on research (see Streuli 2010;

2012) carried out with around 50 children aged 6–14 years, with primary data collected using a variety of qualitative techniques including social mapping, free drawing, and child-led photography; and interviews with parents, teachers, healthcare professionals, and coordinators of *Juntos*. The sample included both recipients and non-recipients of *Juntos* payments.

The fieldwork was carried out in three sites of the Young Lives[2] study in the Department of Ayacucho, in the south-central Sierra of the Peruvian Andes. The sites were selected for their high levels of poverty and because these were also communities covered by the *Juntos* programme. In this paper the communities have been renamed Alamo, Tara, and Vilcas to protect the identities of the research participants. The research took place between 2008 and 2010 and fieldwork lasted about five months. The purpose of the study was to explore, *with* children and the people who were important to them, the everyday implications of a CCT programme. It analysed the role of *Juntos* and related basic services in shaping the everyday lives of children, their families, and their communities. By examining children's views and experiences of *Juntos*, this paper explores the possible role of CCTs in child protection measures.

The situation of Peruvian children

In recent years, Peru has been undergoing a period of political stability and economic growth. Despite this, almost half of the population still live in poverty. Poverty rates only started to fall slightly in the last few years – from 49 per cent to 45 per cent between 2004 and 2006. However, important differences exist within the population. While poverty affects around 37 per cent of the population in Lima, the incidence of poverty in rural areas is around 72 per cent (Instituto Nacional de Estadistica e Informatica, Peru). Peru's total population is 28 million, 30 per cent of which are children under the age of 14 (INEI 2007). Children are the age group most affected by poverty in Peru. While in 2004, 45 per cent of Peru's total population was living in poverty, around 60 per cent of children aged 3 to 16 years were doing so (INEI 2006). The situation was even worse in rural areas where poverty amongst this age group reached 80 per cent.

In the 1990s, the country went through a period of economic adjustment and structural reforms including trade liberalisation. A number of social programmes were implemented to help the most vulnerable population to bear difficulties from the transition. Between 2001 and 2006, important measures were taken to tackle poverty in general, and to improve the situation of children, including, among other relevant initiatives, the development and implementation of: (1) a *Plan Nacional de Acción por la Infancia y la Adolescencia* 2002–10 (National Plan of Action for Childhood and Adolescence), which is based on the United Nations Convention on the Rights of the Child (UNCRC 1989) and sets the strategic priorities of the government with regards to its actions for children; and (2) a *Plan Nacional de Educación Para Todos* 2004 (National Plan of Education for All), approved following the international agreements at Jomtien and Dakar.

In relation to poverty alleviation, two major initiatives were the *Acuerdo Nacional* (National Agreement) and the *Mesa de Concertación de Lucha contra la Pobreza* (MCLCP, or Roundtable for the Fight against Poverty). The *Acuerdo Nacional* is a long-term plan for the economic and social development of Peru agreed by political parties and civil society organisations. Seven out of its 31 long-term state policies are specifically aimed at favouring children. The MCLCP is a multi-sectorial forum including government and civil society, intended to facilitate dialogue and participation in public policies on poverty reduction.

In 2005, the government of Alejandro Toledo implemented the *Programmea Nacional de Apoyo Directo a los Más Pobres* (National Programme of Support to the Poorest), a CCT programme known as *Juntos*, which was continued and expanded by Alan Garcia's government (2006–11), and now also by that of President Ollanta Humala (2011–16).

Juntos: a conditional cash transfer programme

Juntos, like other CCTs, is a programme aimed at poverty reduction. Its main purpose is to build capacities of future generations and break the intergenerational transmission of poverty by giving 'poor' families economic incentives and promoting universal access to education, health, nutrition, and identity services (*Juntos* 2008: 2). The programme claims to follow a basic rights approach, although there is no clear explanation of what this means in practice, and seeks to promote the full participation and oversight of community stakeholders.

Juntos, instituted in 2005, was designed to be distinct from other social programmes that offered monetary transfers without conditions (Francke and Mendoza 2006). Up to the time of the fieldwork, *Juntos* provided a monthly cash transfer of S/. 100 (Nuevos Soles, approximately US$37) to eligible households, which is approximately equal to the average monthly income of families living in extreme poverty and around 17 per cent of the national minimum wage. Initially, the programme was targeted specifically at poor households who had children under the age of 14 (*Juntos* 2007). At the time of the study presented here, the subsidy was the same regardless of the number of children in eligible families. In 2010, *Juntos* started a new phase and some of its activities are now facing important changes.

The cash transfer was mainly given to mothers, but could also be extended to widowers, grandparents, or other caregivers. To receive the monthly cash transfer, recipient families had to enrol their children in school, make sure they attend regularly (at least 85 per cent of the school year) and have the clothing and materials that they needed for school, ensure they get all their vaccinations, and take them for regular healthcare check-ups. Pregnant women had to take part in both antenatal and postnatal care programmes; in addition, adults in the household had to have national identification cards and make sure that their children had birth certificates.

As long as they complied with these conditions, recipient families were eligible to receive the cash transfer for up to eight years; the full S/. 100 a month during the first four years, and a reduced transfer in the final four years, although this was still under review. The cash transfer was suspended for three consecutive months in the first case of non-compliance with its conditions, and indefinitely if non-compliance was repeated. Every three months, *Juntos* local coordinators visited the homes of recipient families to monitor their compliance, and that information was cross-checked with records of school attendance and healthcare visits.

By 2008, *Juntos* had reached 638 of the poorest districts in 14 departments; benefiting a total of 448,787 family households. Identifying recipients was a challenge, given the frequent and profound income fluctuations often experienced by the 'extreme poor'. *Juntos* first targeted specific regions and districts as 'priorities', based on the presence of extreme poverty and inequality, chronic malnutrition, unmet basic needs, and a high level of political violence (*Juntos* 2007). Initially, the programme was seen as a way to tackle the particular vulnerability of populations who were the most affected by the political violence that had been prevalent in the country between 1980 and 2000.

The second targeting stage was at the household level and involved a census of households in the selected districts, using a socio-demographic questionnaire that measures the probability of being a 'poor household' using observable variables such as living conditions, access to public services, and household composition (Francke and Mendoza 2006). Finally, a community validation process takes place, where community and local authorities are brought together, along with representatives from health and education services to ensure that the provisional list of beneficiaries reflects the realities of poverty in the locality.

Juntos also claimed to offer a gendered understanding of vulnerability by directing cash transfers through women since they have generally been seen as more responsible than men in making decisions in the best interests of the entire family (*Juntos* 2007, 2008). Specifically, *Juntos* sought to improve women's bargaining power within the household by reducing their economic dependence and providing them with an independent financial resource.

Implications of *Juntos* upon children's well-being

Children's and parents' views of Juntos

First, the study explored the extent to which *Juntos* was perceived as a programme intended to benefit children. This perception is necessary for the community to have a sense of the programme's accountability to children. Both children and their parents participated in group discussions in which they shared their views and experiences of services related to *Juntos*, such as local schools and healthcare facilities. Within this context, children and parents talked about whether they thought *Juntos* was making a difference in people's lives. Two major themes emerged from the group discussions: the need for increasing the visibility of children within *Juntos*; and the necessity for raising people's awareness about the programme's objectives, strategy, and conditions. Awareness of the programme and its purposes turned out to be limited. Both recipient and non-recipient children, together with their parents, found it difficult to explain the overall purpose of *Juntos*, even though they had been between six and 12 months in the programme. This was surprising among adults, especially women, who received the transfers. Most recipient women found it difficult even to identify *Juntos* conditions they had to meet. Their responses about their obligations under the programme ranged from having to '*sweep streets*' and '*clean the church hall*' to '*taking children to the health centre*' and having to '*buy school materials for their children*'. They seemed unclear about the central idea and purpose of the links between the cash transfer and the conditions they had to fulfil to receive it.

Although all children in the study were to some extent aware of the existence of *Juntos*, they, like the adults, had a limited understanding of its purposes. Most recipient and non-recipient children said the programme aimed to help '*poor families with loads of children*', while others highlighted the role of *Juntos* in supporting children's nutrition and education. However, like their parents, children did not spontaneously associate the programme with a place, an institution, or as being 'for children'. When asked about services, programmes, or places where children and families can get support, parents and children did not mention *Juntos* as one of them. Their immediate responses were that there was '*nothing available*' for them (boy, 9), '*nothing meant for children*' (boy, 13), and '*there is no such thing in our community*' (girl, 12).

When asked specifically about *Juntos*, most recipient children knew that the programme was giving money to their mothers; they even knew the exact amount of the transfer. The children and adults, however, were less clear about the child-specific conditions of the programme such as attending school and having regular visits to the local health centre. Most people interviewed said they did not have to do something special for their children, as most of them saw meeting both the education and health-related conditions as simply requiring something they had been doing anyway before the programme started.

Adults and children from both recipient and non-recipient families viewed *Juntos* as a programme for mothers rather than for children. Overall, children were less 'visible' in people's descriptions of the programme. This does not mean that *Juntos* was not having important effects on children's lives, as will be discussed later; but, as the following quote illustrates, there was a lack of communication about the programme's child-specific aims:

> *It is helping women more than children. Women are now getting better clothes and having their own crops... how much of this goes to children, we don't know. (Nurse, Vilcas)*

Children as seen by Juntos

The way we see people may determine the way we treat them. For this reason, my study explored the images of children and the approaches to childhood, held by both basic services providers and *Juntos* coordinators. In interviews with these adults, they all agreed that they wanted to act in children's best interests. But at the same time, children tended to be portrayed in their narratives either as the 'cause of poverty' for their families, or as 'victims' of lack of care on the part of their families and communities, illustrated in the following quotes:

> *They are poor because they keep having children ... More children, more poverty. Families don't really care for their children. (*Juntos *coordinator, Tara)*

> *The problem with parents in this district is that most of them do not care about their children's health or education. They only want them to help in the fields. They keep having more children and do not look after them properly ... that's why they remain poor. (Primary school teacher, Alamo)*

These statements raise the question of whether service providers held a narrow view of families and childhood living in poverty that they projected onto the community without understanding the very different ideals of childhood that can pertain in different social and cultural settings. The children participating in the study certainly were vulnerable in a number of ways. By attributing that vulnerability to the failure of their families and communities, however, *Juntos* risked failing to address wider social, economic, and political factors such as inequality, social exclusion, and discrimination that contribute to participating children's situation. Such limited understanding of what most affects children's well-being threatens adverse effects in two important ways: it can pathologise families and communities; and it can provide excuses for the State to minimise its role and responsibility for people's welfare.

Children's health

The findings suggest that the families who participated in the study were more frequently using the health services provided in their communities than they had done prior to their inclusion in the programme. These included an increase in visits to clinics by recipient children, in vaccinations, and in ante and postnatal check-ups among recipient women (see also Jones *et al.* 2008). But this usage of health services did not figure prominently in children's responses. When asked about the implications of *Juntos* upon their lives, recipient children said the programme was having a positive impact on their families' economic and food security. Some of them said that they now '*get more food*' and that their mothers were '*using the money to improve the house*' or to '*buy more materials for school*' as illustrated in the quotes below:

> Interviewer Do you think Juntos is making any difference in families' lives?
> *Boy 1 Yes, we are better now.*

Interviewer How come?
Boy 1 In the money, food, and clothes too.
Boy 2 My mum buys me clothes, and for her as well. She also bought for my dad. She buys food too, such as grapes, fruit, and peaches.
Boy 1 Yes, fruit, and chicken to kill [and eat]. Things to store as well [cupboards, organisers] (Recipient boys, six and seven years old, Alamo)

Children were also invited to discuss what they thought mothers did with the money given by *Juntos* and explore how the subsidy was spent within the household. Consistently, recipient children in the three sites said that their mothers spent the money to '*cook*', or to buy food items like milk, fruits, and other things that they cannot produce themselves, which they used to '*feed their children*'. The following quote from a recipient child illustrates this:

Interviewer And what do mums do with the cash?
Boy They buy meat, milk, and cheese.
Interviewer For whom?
Boy For all of us. We eat. Sometimes she brings fruit for us to eat, to feed us. Cheese as well. There is no cheese here (in the community), that's why she goes there (to Ayacucho, main city)... Sometimes we have cheese here, but we only have a small cow and it gives little milk, so we only drink it. (Recipient boy, nine years old, Tara)

Children, however, also highlighted that even when their families are getting more food items, these may not be equally distributed among different members within the household. For example, a recipient girl pointed out that her mother '*always keeps part of* [the girls'] *food for her grandma who comes tired from working in the fields*', while another recipient girl, aged seven, said, '*My mum buys us fruits, food, but sometimes my older brother gets more because he goes to the city to work*'.

Education and schooling

Interviews with children, their parents, and teachers suggest that *Juntos* is having a positive effect on school enrolment and attendance rates for primary school (see also Jones *et al.* 2008). For example, a six-year-old recipient girl explained, '*My mum is now making* [me] *read and study more* [and] *I now got all my school materials to do homework*'. Since the programme 'punishes' those families who do not send their children to school regularly, the majority of families in the area are now taking more responsibility in relation to education. According to a father, '[They] *didn't send them before* [to school] ... [they] *didn't see the value* ... [they] *sent them to the fields to help instead*'.

In particular, since *Juntos* started, more children with special needs or with either physical or cognitive disabilities are being taken to school. This is having a positive impact on the children and their families, as the former can now benefit from education and the company of peers, and the latter spend less time caring for them and can go to work in the fields or in the city. However, teachers also complained that they are not properly trained for these special cases, nor have they the appropriate educational materials. Therefore, the extent to which these children are benefiting from schooling remains uncertain. Partly as a result of the lack of resources and of teacher training for children with special needs, some of these children are experiencing social exclusion and violence within the school setting, as explained by one of the mothers whose 13-year-old son is deaf, and cannot talk:

> *It is good that he goes to school. I didn't send him before, because they [the school] didn't want to take him. Now with* Juntos, *all children have to be in schools. But now he cries*

> *every time he has to go [to school] . . . I didn't know why. The boy next door told me that children tease him during recess and that his teacher beats him when he doesn't stay quiet. (Recipient mother, Alamo)*

Despite the advantage of poor children getting to school, some recipient children and families were experiencing stigmatisation and discrimination by teachers and providers of other basic services for not complying with *Juntos* conditions. Some mothers in the programme complained that they were singled out for specific failures – such as not having bought school uniforms – during monthly meetings organised by *Juntos* coordinators. Their children singled out faults of their schoolmates who were not recipients with such phrases as '*he hasn't got his vaccinations*', '*she never has pencils and eraser*', '*he looks untidy and dirty all the time*'. These comments suggest status tensions between recipients and non-recipients around the obligations of CCT recipients. A recipient mother said, '*Juntos is everywhere, we need to be careful and comply with what they say, otherwise we lose the money*'. This surveillance tactic held by the programme's stakeholders is also shown in the next quote:

> *Sometimes we pick a classroom at random and check that recipient children have their notebooks, that they are clean and neat, that they have pencils, and their books covered in plastic wrap. Children know that and they try to get everything ready. Some of them get upset with their mums if they are pointed out by* Juntos *for not having all their materials. (*Juntos *promoter, Vilcas)*

Both children and parents mentioned other difficulties with schooling, such as low quality teaching and poor school facilities. Most of the parents interviewed felt they could do little to help their children with schoolwork since they do not feel prepared and do not have enough time. Children also said they would like to have access to a library or places where they can do their homework and learn new things. Parents and children also mentioned the difficulties that most children in the area have to continue their education beyond primary school. Most of them have to travel long distances to secondary schools. Children usually spend hours on the highway waiting for public buses to take them to school because these prefer to pick up adult passengers, whose fare is nearly five times higher than that of students. Parents also expressed concerns about the future of their children once they have finished school, as there are few jobs available in the area.

As pointed out earlier, CCTs are very effective in getting and keeping children in school. Being in school, however, raises further problems for disadvantaged children, which are not normally addressed in such programmes, such quality of schooling, their everyday experiences of school, and their entry to the labour market. These issues were identified by both adults and children as main concerns in relation to education in the area. The CCT could get children into school, but it could not ensure that the schools themselves were any more accessible or worth attending than they had been before.

Other aspects of children's well-being

Questions were raised that challenged the wisdom of establishing *Juntos* as a stand-alone programme rather than as an element in a more comprehensive system. In general terms, parents valued *Juntos*' intervention on health and education, but they indicated they would also like the government to address other child-specific vulnerabilities that they identified during group discussions, including the prevalence of respiratory diseases, the impact of environmental shocks (such as hail and frost) on children's health; access to services aimed at children with special needs; and children's experiences of violence within the family, school, and community. Even though *Juntos* aims to provide support to families with children, most of the parents who participated in the study felt they could not get any assistance from the programme in case of need. It was not expected that *Juntos* would be responsible for all problems faced by children and parents, but the programme should be connected to, and be part of, a more sound and functional social protection system. It should be complemented by child-specific services and programmes to which parents could be referred when needed.

In particular, parents said they would like *Juntos* to be more aware of and provide for children and families in special circumstances. For example, one of the participant families who had a child with physical disability reported that the cash transfer was not enough. Their child is now attending a mainstream school, but is not getting the attention and care she requires. They need a specialised school, as well as specialised healthcare, which is not available in the communities studied and would be extremely costly to access in the nearby city.

Some pointed out a need to recognise family differences, such as when extended family is caring for the children. During group discussions in Tara, for instance, women commented that grandparents looking after their grandchildren find it difficult to comply with certain conditions of *Juntos*. Likewise, according to a group of non-recipient parents in Vilcas, *Juntos* does not always sufficiently consider the most vulnerable. For example, they mentioned the case of elderly people who are getting a pension from the government. If the amount of money exceeds S/. 100, they are not entitled to receive the cash transfer, despite the number of children they are looking after.

There were also worries about the long term. Both individual and group discussions with children and parents revealed uncertainty about how long *Juntos* will last. Apparently, families have not been informed when the programme will end, which raises important questions in relation to its sustainability. So far, recipient families are not being supported to develop strategies that could help them sustain themselves after *Juntos*. One sees this in concerns that arose regarding long-term prospects for children's education. How will children transition to secondary school when their families 'graduate' from the programme? Most children said they would like to continue their studies, even to become professionals, but their families had a more immediate concern about what would happen after they finish primary school.

Reflections and recommendations

This paper raises the question of whether CCTs could play a greater role in child protection than is now the case. The answer is yes, they can and they should. Investing in child protection under the umbrella of social protection could be one of the most efficient and sustainable strategies to tackle poverty (Crawford 2001). This study of *Juntos* suggests that it, like other CCTs, is having important effects on certain aspects of children's well-being such as physical health, nutrition, and schooling. At the same time, however, there are limitations to these benefits, occasioned by the fact that the CCT cannot guarantee the quality and usefulness of government services that it obliges recipients to patronise. Given that a certain amount of stigmatisation seems to accompany recipient children and families, one might also ask questions about effects on other aspects of children's well-being and development, such as individual self-esteem, gender-based notions and practices, cultural identity, and relationships within the community.

Taken together, the study findings suggest a cultural distance between the CCT and the people it is intended to serve that may explain the observed lack of communication, the lack of attention to quality of referred services, the misunderstanding of family roles, and other factors limiting the programme's benefits. It is clear that the way children and their families were perceived by the programme and its staff influenced how they were treated. In a way, recipient women and children could not win. On the one hand, the programme holds a traditional view of women that places on them the entire responsibility of the household and children's care. Men are scarcely involved. But the CCT then traps the women with expectations apparently not realistic for their situation. The overall image of children that *Juntos* was promoting at the time of the fieldwork reflected normative ideas about childhood and child development mainly coming from outside the community, perhaps even from industrialised countries. This is illustrated by failure to attend to the needs of grandparents and pensioners who in much of the community are the ones actually in charge of children. The programme associates normative childhood with dependency, schooling, and nuclear family structures not always readily accessible in the communities studied. This makes children and their families seem less capable, less motivated, and less 'developed'. For this reason, teachers and *Juntos*' promoters assumed that recipient children and their families needed constant monitoring and direction from them on how to behave and to meet their CCT obligations, and one could interpret the findings to suggest that the CCT may have tried to use the programme's conditions to change recipients' attitudes and behaviour less to what was needed than to match professionalised assumptions and ideas of what kind of childhood and family life is normal. One suspects from the data that many children may in their relations with *Juntos* have to deal with unjustifiably narrow views of who they are and accept how others want them to live.

The findings also suggest that a narrow view of childhood as potential human capital may induce *Juntos* to retain a rather traditional approach to childhood poverty in a situation in which more flexibility and innovation might have been helpful. An overemphasis on children's future skills and contributions as adults diverts attention, as noted above, from present needs and experience while they are still children. One could also question the assumption that what is good for the household will automatically be good for children. *Juntos*, like other CCTs, believes that cash transfers to the family household, and access to education and health care for the children, are the main means to improve the living conditions of people in poverty. But is that the case for an eight-year period, one that does not see needy children into secondary or tertiary education? While the programme over even this limited period may help some people, what is the effect of other problems beyond access to monetary resources and basic services, such as social exclusion, stigma, and discrimination, and experiences of violence that the CCT also seemed to engender?

Despite its innovative design in some respects, *Juntos* is in others still reproducing paternalistic approaches to development in which the programme dictates behaviours at the family and individual level, and in which there is relatively little choice and little participation by recipients in decisions. There is no structured place where they can voice ideas and concerns that the programme will hear, learn from, and react to. One can ask what kind of individual and social development arises from that sort of organisation.

For all these reasons, CCTs like *Juntos* should go beyond just raising income and consumption standards of people living in poverty, as helpful and important as that is, and become more 'transformative' by enhancing social equity and social rights of the people involved (Devereux and Sabates-Wheeler 2004). These programmes should incorporate a rights-based approach that recognises the need for integrated, inter-sectorial approaches to child-related policies, which would help with the problem of CCTs as stand-alone programmes referring people to services that may be substandard. Such an approach would involve gathering information that goes beyond health and education statistics, and includes data on child abuse and exploitation, as well as civil and political rights of children. It certainly would include systematic dialogue with both children and their families.

To be more child-sensitive, CCTs need to consider children within the context of their relationships with their families and wider society (Ray and Carter 2007). Some of these children experience systematic and institutionalised exclusion, and for this reason, it is crucial to understand the roles of different actors and institutions in the lives of the children. More children-centred CCTs also would improve knowledge and awareness of child-specific risks and incorporate different family models and structures, including grandparents, the extended family, and child-headed households. They would also pay more attention to community history, backgrounds, and social cohesion and dynamics in order to minimise tensions and maximise people's use of social networks. CCTs like *Juntos* need to combine their focus on individual households with community-level actions so as to strengthen mutual collaboration and sense of reciprocity, which is vital to most rural communities in the country. Finally, to be more child-sensitive, CCTs should build better links between child protection and broader social protection measures.

Acknowledgements

The study presented in this paper was done as part of my doctoral work (Streuli 2010) in Childhood Studies at the Institute of Education, University of London, and as a collaborative effort with Young Lives, a 15-year longitudinal study of childhood poverty in Ethiopia, India, Peru and Vietnam (www.younglives.org.uk).

Notes

1. This paper is published posthumously. Because it contained information and ideas that are important to this issue, the original draft was revised by the guest editors, who took into account comments by an anonymous reviewer, with the help of Virginia Morrow. They hope they have been true to Natalia's intentions.
2. For information about Young Lives, see the Preface to this issue.

References

Bouillon, C. and L. Tejerina (2006) *Do We Know What Works? A Systematic Review of Impact Evaluations of Social Programmes in Latin America and the Caribbean*, Washington, D.C. Inter-American Development Bank.

de la Brière, B. and L. Rawlings (2006) 'Examining Conditional Cash Transfer Programmes: A Role for Increased Social Inclusion?', available at: http://siteresources.worldbank.org/SOCIALPROTECTION/Resources/SP-Discussion-papers/Safety-Nets-DP/0603.pdf (retrieved 20 February 2007).

Crawford, P. (2001) 'Child protection: theoretical background', in I. Ortiz (ed.) *Social Protection in Asia and the Pacific*, Manila: Asian Development Bank.

Devereux, S. and R. Sabates-Wheeler (2004) *Transformative Social Protection*, Brighton, UK: Institute of Development Studies.

Francke, P. and A. Mendoza (2006) 'Perú: Programmea Juntos', in E. Cohen and R. Franco (eds.) *Transferencias con Corresponsabilidad. Una Mirada Latinoamericana*, Mexico City: FLACSO Mexico – SEDESOL.

Gordon, D., S. Nandy, C. Pantazis, S. Pemberton, and P. Townsend (2003) *Child Poverty in the Developing World*, Bristol, UK: The Policy Press.

INEI (2006) *Condiciones de Vida en el Perú: Evolución 1997–2004*, Lima: Instituto Nacional de Estadística e Informática.

INEI (2007) 'Censo Nacional de Poblacion y Vivienda 2007', available at: www.inei.gob.pe/ (retrieved 22 January 2008).

Jones, N., E. Villar, and R. Vargas (2008) 'Cash transfers to tackle childhood poverty and vulnerability: an analysis of Peru's *Juntos* programme', *Environment and Urbanization* 20 (1): 255–73.

Juntos (2007) '*Juntos*: legal framework', available at: www.juntos.gob.pe/intro.php (retrieved 20 December 2009).

Juntos (2008) *Informe de Evaluacion Anual 2007*, Lima: Programmea Nacional de Apoyo Directo a Los Mas Pobres – '*Juntos*'.

Morley, S. and D. Coady (2003) *From Social Assistance to Social Development. Targeted Education Subsidies in Developing Countries*, Washington, DC: Center for Global Development.

Ray, P. and S. Carter (2007) *Understanding and Working with Children in the Poorest and Most Difficult Situations*, London, UK: PLAN UK.

Streuli, N. (2010) 'A Study of How Peruvian Children Involved in a Social Protection Programme Experience Wellbeing and Poverty', unpublished PhD thesis, London, UK: Institute of Education, University of London.

Streuli, N. (2012) Children's experiences of *Juntos*, a conditional cash transfer scheme in Peru. Working Paper 78. Oxford: Young Lives.

UNICEF (2005) *The State of the World's Children 2006: Excluded and Invisible*, New York: United Nations Children's Fund.

Valencia, E. (2008) 'Conditional cash transfers as social policy in Latin America: an assessment of their contributions and limitations', *Annual Review of Sociology* 34: 475–99.

Listening to Iraqi refugee children in Jordan, but then what? Exploring the impact of participatory research with children

Martha Nelems and Vanessa Currie

Following the American-led invasion of Iraq, thousands of Iraqis fled to Jordan and the international donor community initiated humanitarian assistance. Through a unique partnership, three organisations conducted participatory research with Iraqi children and their families in Amman. The goal was to understand children's lived experiences – their challenges and coping strategies – with a specific focus on child protection. A better understanding of local context had an immediate, positive impact on organisations and their effectiveness, but long-term change proved elusive. The authors explore the potential for participatory research to transform programming and the obstacles to institutionalising change.

Écouter les enfants réfugiés irakiens en Jordanie, mais après ? Explorer l'impact de la recherche participative avec les enfants

Après l'invasion de l'Irak lancée par les États-Unis, des milliers d'Irakiens se sont enfuis en Jordanie et la communauté internationale des bailleurs de fonds a mis en route une assistance humanitaire. À travers un partenariat unique, trois organisations ont mené des recherches participatives avec les enfants irakiens et leurs familles respectives à Amman. Le but était de comprendre les expériences vécues par les enfants – leurs défis et leurs stratégies d'adaptation – en se concentrant précisément sur la protection des enfants. Une meilleure compréhension du contexte local a eu un impact immédiat et positif sur les organisations et leur efficacité, mais le changement à long terme s'est révélé difficile à atteindre. Les auteurs examinent le potentiel pour la recherche participative de transformer la programmation et les obstacles à l'institutionnalisation du changement.

Ouvindo as crianças refugiadas do Iraque na Jordânia, mas e depois? Explorando o impacto da pesquisa participativa com crianças

Após a invasão do Iraque coordenada pelos norte-americanos, milhares de iraquianos fugiram para a Jordânia e a comunidade doadora internacional iniciou a assistência humanitária. Através de uma parceria exclusiva, três organizações conduziram uma pesquisa participativa com crianças iraquianas e suas famílias em Amman. O objetivo era compreender as experiências vividas das crianças – seus desafios e estratégias de lidar com a situação – com um enfoque específico na proteção infantil. Uma melhor compreensão do contexto local teve um impacto imediato e positivo sobre as organizações e sua efetividade, mas a mudança no longo prazo mostrou-se difícil. As autoras exploram o potencial para a pesquisa participativa transformar os programas adotados e os obstáculos para institucionalizar as mudanças.

¿Qué sigue tras escuchar a los niños y niñas iraquíes refugiados en Jordania? Análisis de los resultados de una investigación participativa con niños y niñas
Tras la invasión de Estados Unidos a Irak, miles de iraquíes se refugiaron en Jordania y la comunidad internacional de donantes comenzó a proporcionar ayuda humanitaria. Tres organizaciones concertaron un acuerdo específico para realizar una investigación participativa con niños y niñas iraquíes y sus familias en Amán. El objetivo era comprender las vivencias de la niñez, sus desafíos y qué hicieron para enfrentarlos, centrándose en la protección de la niñez. El hecho de comprender mejor el contexto local repercutió de forma inmediata y positiva en las organizaciones y en la eficacia de su trabajo, pero no se produjeron cambios a largo plazo. Los autores examinan el potencial de la investigación participativa para la transformación de programas y los obstáculos que existen para transformar las instituciones.

Introduction

With the 2003 invasion of Iraq by American-led coalition forces, thousands of Iraqi families fled to Jordan. While some of the wealthier families joined the established Iraqi community in West Amman, the majority moved into the poorer parts of East Amman. Iraqi children and adults lived through many difficult experiences during the war in Iraq, including personal experiences of violence, but they also faced new challenges relating to their life in Amman: economic hardship, discrimination, violence, and social, cultural and religious isolation.

The initial efforts of the international donor community to support Iraqis in Jordan focused on basic healthcare and other forms of humanitarian assistance. However, access to education for Iraqi children soon attracted the attention of the donor community and many engaged in high-level advocacy with the Jordanian government. By August 2007, thousands of Iraqi girls and boys were registered in Jordanian schools.[1] With this achievement, international non-governmental organisations (INGOs) and UNICEF began to shift their focus to the more elusive issue of 'child protection for Iraqi children'. Largely as a result of the advocacy work of Queen Rania, child protection in Jordan was primarily associated with 'child abuse', rather than a broader understanding of child protection as the prevention and response to abuse, neglect, and exploitation of children and violence against them. 'Psychosocial' interventions were deemed one of the best ways to address child protection problems – though there was little understanding of what psychosocial support for the Iraqi community should look like.

In October 2007, a group of three organisations undertook an assessment of the situation of Iraqi children living in Amman through participatory action research. The Unity Circle Project involved collaboration between the International Institute for Child Rights and Development (IICRD) at the Canadian-based University of Victoria and two INGOs working in Jordan, Save the Children (SC)[2] and Relief International-Schools Online (RI-SOL). The project was conducted with the financial support of the Canadian International Development Agency (CIDA). The UNICEF office in Jordan provided financial support for developing the final report and organised its distribution after the public launch in October 2008.

Researchers developed three key questions to be answered during the participatory process with Iraqi children:

- What are the challenges to well-being facing Iraqi children in Amman?
- What coping strategies are Iraqi girls and boys using to respond to these challenges?
- What positive coping mechanisms or strengths identified by Iraqi children and their families can be encouraged and built upon in developing and improving programming interventions?

These questions derive from IICRD's commitment to build on strengths and to help empower local communities to develop innovative, local solutions to self-identified challenges. IICRD's strengths-based approach draws from rights-based theory, but is also based on evidence. It was developed by IICRD after considerable practical experience with children globally, as well as extensive literature reviews, both of which affirmed the developmental importance of children and their families actively participating in identifying and solving problems.

Over several months, these research questions were explored through experiential activities, such as community mapping, journey drawings, poetry, and focus group discussions. Ninety-six Iraqi girls and boys, aged 5–17, met in six different groups in East Amman, and one group in West Amman. Researchers also met with the siblings and parents of participants whenever possible.[3]

The Unity Circle project enabled Iraqi girls and boys to describe the challenges they experienced both in Iraq and Jordan in their own words and images, as well as identifying strengths at the personal, community and cultural levels. A participatory analysis of the findings followed, shaping the development of the final report. The report documented all of these components and recommended general principles for programming with Iraqi children in Jordan as well as making specific recommendations in the education and psychosocial sectors.

Advances have been made in the development of tools and techniques for consulting children about their lives and the programmes intended to support them. However, questions remain as to whether and how organisations make use of the information from children and families to improve their policies and programmes. These questions have received surprisingly little attention in the literature despite their critical importance. Although organisations operate under constraints that hinder their ability to adapt programming to new information gained through participatory processes, opportunities also exist. But what are these constraints and opportunities, and is it realistic to expect learning from processes such as participatory research to guide an organisation's policy and programming? Through a series of interviews with key players in the Unity Circle project, the authors explore these questions.[4] They investigate what happened after the project ended, and whether the participating organisations succeeded in their efforts to take action on the project's findings.

Research findings and recommendations

Iraqi girls and boys identified a wide range of challenges related to their experiences during the war in Iraq, including the death of parents and other loved ones, missing relatives, and witnessing the destruction of their homes and communities, and/or violent events including murder. Some Iraqi boys shared their experience of being kidnapped and held in captivity. Many of the Iraqi children shared their feelings of helplessness at being unable to protect their families, a significant loss given the cultural value Iraqis place on bravery and courage.

Iraqi children discussed many challenges, which related to their displacement from Iraq, starting with what many described as an arduous journey, and grief at the loss of their physical and social environments. These losses pertained to every area of their lives: peer and family networks, places of worship, financial security, and – importantly for children – gardens and recreational opportunities.

Their new 'home' in Jordan posed challenges to all of the Iraqi children living in East Amman, from danger related to traffic, to overt discrimination and violence experienced both in their neighbourhoods and schools. There were many education-related problems for Iraqi children in addition to violence, including difficulty understanding the Jordanian dialect of Arabic and the new school curriculum, poor quality of teaching, and in some cases, the inability to attend school. However, the violence and discrimination experienced by Iraqi girls and boys in East Amman neighbourhoods and schools was the issue that most dominated discussions throughout the project.

Iraqi youth aged 14–17 raised additional issues such as their frustration at being unable to work and contribute to family income, a particularly acute problem with so many fathers and male relatives missing in Iraq. Some Iraqi youth were unable to attend school due to care responsibilities at home, and others found it demoralising and painful to be in classrooms with much younger cohorts as a result of missed years of schooling during the war. The lack of privacy in cramped living conditions in Amman was also identified as a problem by Iraqi youth.

Some key differences were observed between children living in the poorer (East) and wealthier (West) districts of Amman. Iraqi children in West Amman found their schools generally supportive environments and did not experience discrimination and violence in their neighbourhoods, at least not on a regular basis. Although Iraqi children in West Amman shared with their East Amman peers ongoing anxiety about missing relatives and male family members working in Iraq, they typically had greater financial security.

While Iraqi girls and boys experienced many similar challenges, they observed key differences based on gender. Although almost all children felt helpless to protect their families, this feeling was particularly acute amongst Iraqi boys, especially those boys who were unable to protect their mothers and sisters from witnessing violence, or from being questioned by American forces. In the Unity Circle project, kidnapping was experienced exclusively by Iraqi boys. Both girls and boys recognised that parental over-protectiveness was particularly intense for girls,[5] contributing to girls' greater sense of isolation and boredom. This lack of freedom also contributed to girls' inability to participate in religious life. While most children participated in far fewer religious ceremonies in Amman than previously in Iraq, girls had even less access to religious buildings than boys, with some girls never visiting a mosque or a church in Amman. Violence in the schools and at times on the streets, was experienced by both girls and boys, though in different forms: Iraqi boys talked about being punched and kicked by Jordanian boys, while Iraqi girls reported their most frequent violent experience to be hair pulling.

Iraqi parents expressed worry about many of the same problems as their children. Parents also identified specific challenges that they faced in their role as caregivers, including not feeling empowered to teach children important cultural values and appropriate behaviour, over-protectiveness, and concern about the consequences of lying to their children. Some Iraqi parents admitted to the use of violence in their home against their children.

Researchers learnt how Iraqi girls and boys pro-actively tried to solve their problems or mitigate the impact of them. A range of their coping mechanisms were identified. 'Positive' coping strategies were: cooperating and spending time with families; contributing to household income; participating in activities at community centres; integrating into Jordanian life; helping others and engaging in problem solving, especially with anxious or depressed parents; building friendships with other Iraqis and Jordanians; and maintaining relationships with family via technology. 'Positive' values were being optimistic about the future and being patriotic about Iraq. Youth identified additional positive coping strategies such as exercising, painting or writing poetry, negotiating with parents for greater freedom, avoiding asking their parents for help, seeking solitude, and protecting other Iraqi children, for example by walking younger children to school to protect them from violence.

Caution was identified by almost all Iraqi girls and boys as a negative coping mechanism, even though some children admitted to being cautious in their new life in Amman. Iraqi youth identified emotional withdrawal, pessimism about the future, and a fixation on past events as emotional strategies that were negative and adversely affected their well-being.

Iraqi children and their parents disagreed on whether some coping strategies were positive or negative. For example, the adoption of a Jordanian accent in public spaces was considered by many Iraqi children to be positive, whereas Iraqi parents were uniformly dismayed and feared that their children were ashamed to be Iraqi. Lying, the use of force, and refusing to talk to their parents about difficult emotions were other coping strategies that were contentious between Iraqi children and their parents.

Many strengths were identified at the personal, cultural and community level, strengths that could be built upon and utilised to ensure more effective programming for Iraqi children. These ranged from participating in Iraqi cultural activities and promoting cultural values such as bravery and generosity, to using the talents of the many educated and unemployed Iraqi adults for the benefit of children, especially in an educational context.

The project's recommended, general principles for programming with Iraqi children were:

- listen and talk to Iraqi children;
- build on Iraqi children's strengths; and
- involve Iraqi parents.

The specific recommendations for the Education Sector were:

- address violence;
- address discrimination; and
- work to prevent school drop-out and support enrolment.

And for the psychosocial sector:

- support Iraqi children to process difficult experiences in both Iraq and Jordan;
- support the strengthening of family, cultural and religious bonds;
- contribute to Iraqi children's sense of agency;
- support the development of peer networks;
- build opportunities for children to exercise, be artistically creative and access nature;
- ensure safety;
- create youth-specific programming;
- support Iraqi adult volunteers; and
- build local capacity.

After analysing the project findings with researchers, each group of children chose a specific problem they wanted to address and with supportive adults, they developed and implemented an action plan to address this challenge. Action plans built upon the strengths identified by children, ensuring that ownership remained with the children themselves. One group developed a sports and recreational programme to address their boredom and isolation in Amman, while another group held a special cultural day at their community centre where they shared their pride in being Iraqi with other Iraqis and Jordanians.

How understanding children's perspectives improved organisations' effectiveness

How did learning from Iraqi children about the challenges to their well-being and their coping mechanisms and strengths help to improve the effectiveness of the organisations involved? Six significant improvements in performance were identified by the organisations involved in the Unity Circle project:

- willingness to increase public awareness about the challenges facing Iraqi children;
- improved relationships with Iraqi adult volunteers;
- new-found confidence to engage in child protection programming;
- prioritised child protection interventions based on Iraqi children's perspectives;
- refocused attention on the systemic causes of child protection problems; and
- greater commitment to participatory development in other areas of work, in addition to children's programming.

Making public Iraqi children's challenges

Participating organisations all share the perspective that they learnt about the reality of Iraqi children's lives, and as a result of 'the sound evidence' they helped to generate, they were better equipped to publically discuss challenges experienced by Iraqis in Jordan.

Former deputy director for Emergency Programmes at SC argues that the Unity Circle project '*concretized child protection needs … and made it hard to refute the problems Iraqi kids were having*' (Jason Erb, Former Deputy Director for Emergency Programmes Save the Children; Interview). IICRD authors confirm that the new data was welcomed by many members of the international donor community, and was used as the basis for initial needs assessments. For example, AusAid reported using the project report as the starting point for their needs assessment of Iraqis in Jordan. Some senior members of the international development community in Jordan reported that the project findings were the first time they heard Iraqi voices articulate problems within their community, a surprising situation given the physical proximity of Iraqi refugees in East Amman to embassies and the offices of INGOs and UN agencies in West Amman.

As a result of having better data on the reality of children's lives, organisations reported changing their advocacy approach with the Jordanian government. According to the then director of the smaller INGO partner in the project, RI-SOL, the Unity Circle results helped them to '*connect with authorities and start a debate*' (Angelita Carreda, Former Country Director, RI-SOl; Interview). Whereas SC's advocacy work was previously based somewhat on assumptions, they used the data about Iraqi children to improve the credibility of their case that special attention needed to be focused on child protection programming. SC began to raise the issue of violence in the schools systematically in education coordinating committee meetings. According to SC's child protection specialist, it was a combination of their work and the advocacy work of Queen Rania that led the Jordanian Ministry of Education not only to admit that violence in schools was a real problem but also to be willing to listen to Jordanian and Iraqi children speak to them directly about the problem.

'Going public' on a highly sensitive issue such as violence and discrimination against Iraqi children in Jordanian schools was a real change in approach for some organisations, who had felt unable to do so in the absence of data. At that time, many in the donor community also feared the Jordanian government and viewed them as a host they were unwilling to criticise. As the former director of RI-SOL explains:

> *The launch of the report was really important. It was the first occasion when we were talking publically about some of the problems. Members of the Ministry were there and it was done in a very formal way. It was good. It was not hidden as other things had*

been ... the important point was that we ... came out on problems (Angelita Carreda, Former Country Director, RI-SOL; Interview).

New relationship with Iraqi adult volunteers

A greater understanding of local context not only gave new data to the organisations involved, but in one organisation changed the relationship between staff and Iraqi adult volunteers. RI-SOL credits the Unity Circle with helping them work with Iraqi volunteers in new, more egalitarian ways, based on a greater understanding of the complexity of their lives. According to Karishma Budhdev, Former Deputy Director, RI-SOL,

> *We had to respect the volunteers and I think they benefitted by having professionals talk to them as equals not as beneficiaries ... It impacted relationships. It helped us ask questions of Iraqi volunteers and helped us to start thinking about relationships as opposed to a matter of productivity and reaching results with donors (Karishma Budhdev, Former Deputy Director, RI-SOL; Interview).*

For the deputy director of RI-SOL, the project's greatest impact on their organisation was improved staff relationships with Iraqi volunteers.

Confidence to engage in child protection programming

Both INGOs report developing greater confidence in their organisation's ability to engage in child protection programming: one organisation became aware of their growing expertise in child protection compared with other INGOs, and another expanded its work in child protection. For RI-SOL, being a part of a process that generated new data about critical child protection issues for Iraqi children legitimised their entry into broader child protection programming. This was significant in a context where Queen Rania and her Royal non-governmental organisation (RNGO) the Jordan River Foundation 'owned' the issue of child protection, understood narrowly as child abuse. The process of building confidence was two-fold for RI-SOL. First, RI-SOL was glad to be working with IICRD researchers, whom they viewed as specialists in child protection. Second, direct involvement in the Unity Circle project encouraged senior staff at RI-SOL because they were growing in expertise on children's issues and could develop relevant programming for children beyond education. The deputy director argues that their involvement in the project helped them

> *justify our asking UNHCR for more money and to do more than education. It gave us a reason for extension beyond informal education and to see additional activities with communities (Karishma Budhdev, Former Deputy Director, RI-SOL; Interview).*

The trend of gaining confidence and building legitimacy in order to build child protection programming was also true for SC. The former deputy director maintains that they had not focused on child protection before partly as a result of the Jordan River Foundation's dominance on the issue, but also due to a lack of clear thinking about child protection. According to then deputy director of SC,

> *The Unity Circle report happened at a critical time and helped crystalize some of the thinking that had been going on. At a minimum, it helped accelerate our thought process and helped us articulate psychosocial support as part of a larger child protection approach in the public schools. It put us on the map on child protection in a way we wouldn't*

have been before... (Jason Erb, Former Deputy Director of Emergency Programmes; Save the Children; Interview).

Prioritising interventions based on Iraqi children's perspectives

Both INGOs and UNICEF reportedly re-structured their interventions to engage in child protection programming that addressed Iraqi children's priorities. Thus, an important part of the participatory action research process was the prioritisation of issues by children during their analysis of the data with adult researchers and INGOs.

New information about Iraqi children was the basis on which SC staff determined their child protection programmes, and they prioritised within this portfolio according to Iraqi children's assessment of the most critical issues. As a result, SC shifted their focus from helping Iraqi girls and boys in accessing schools, to addressing violence and discrimination in the schools. SC's child protection specialist argues that, as a direct result of the Unity Circle project,

> *We created the help desk and networked with other service providers; we did proper case management; we hired social workers; we refocused on 'do no harm' and 'best interest of the child' principles; we worked with teachers and principals; and we advocated at the Ministry for Iraqi families (Nadine Haddad, Child Protection Specialist, Save the Children; Interview).*

Other senior staff at SC argue that the Unity Circle did more than help them prioritise within the child protection envelope: it refocused them on the child. Having seen the impact of incorporating Iraqi children's perspectives in their work, SC believes they put Iraqi children more at the centre of their programming. The then deputy director explains that the project '*was a watershed moment in that it took us over the hump so we could continue to build on other child-led work and experiences in the region and push it further at a practical field level.*'

According to RI-SOL's former director, their organisation prioritised the Unity Circle's recommendations by using them as the basis for their annual proposal to the United Nations High Commissioner for Refugees (UNHCR). In fact, the Unity Circle recommendations shaped RI-SOL's entire programming focus.

UNICEF argues that it also prioritised new child protection programming on the basis of the Unity Circle results. According to UNICEF's emergency education officer, UNICEF '*relies on researchers and surveys to develop new projects as well as our observations*' (Deema Jarrar, Emergency Education Officer, UNICEF; Interview), and the Unity Circle recommendations were part of the decision to embark on an ambitious programme to educate all school counsellors in Jordan about violence in the schools.

Refocusing on the systemic causes of child protection problems

An unexpected but significant outcome of the Unity Circle project for SC was how Iraqi children's insistence on the centrality of violence in schools led to a profound shift from psychosocial interventions for Iraqi children to the broader rubric of child protection for all children in Jordanian schools – Iraqi, Jordanian, and others. According to SC, this programmatic evolution was well received by the Jordanian government, long worried about the societal impact of programming for Iraqi children that excluded poor Jordanian children. It also resulted in senior SC staff thinking more strategically about the systemic causes of child protection problems. Although SC's former deputy director does not attribute this development directly to the project, he believes that conversations that began with this project contributed to the organisation moving away from focusing on Iraqis as a special consideration and seeing them instead as part of a larger development context in Jordan. For the former deputy director, the great 'advocacy win' of the Unity Circle project was a deepened understanding on the part of many in the international development community that '*to lift the boat for Iraqis, you had to lift the boat for Jordanians as well ... They had the same problems as Iraqis ... you couldn't disentangle their issues from the other issues*' (Jason Erb, Former Deputy Director of Emergency Programmes, Save the Children; Interview).

Commitment to participatory development

One of the most profound, and another entirely unexpected, outcome of the project was how organisations became convinced of the intrinsic value of participatory engagements with beneficiaries to understand the local context better. Being involved in the Unity Circle project built commitment on the part of all the organisations involved to participatory engagement in general, not only with children, but in other programming as well. The participatory action research process with Iraqi children also had a significant impact on RI-SOL senior staff at a personal level. For both the former director and deputy director of RI-SOL, learning participatory approaches was very important for their organisation at the time and proved that research can have practical applications:

> *It [the Unity Circle project] had tremendous impact. Mentoring staff and helping them think in new ways ... A lot of good exercises were learned about participation. We had not been involved with beneficiaries in the same way before. We started to involve kids, communities, organisers, families and it [new programming] was Unity Circle on a different level Karishma Budhdev, Former Deputy Director, RI-SOL; Interview.*

While they were still at RI-SOL, these two senior leaders began designing all of their programming differently to include approaches that were more participatory. Both of these former senior staff argue that commitment to participatory development has remained a core value in their work, long after they have left RI-SOL.

Although engagement in participatory research was less of a revolutionary experience for SC, it confirmed for them that adopting participatory approaches was useful and validated the inclusion of participatory approaches within their work. Subsequent to the Unity Circle project, SC used the well-known 'Child to Child' methodology in Jordanian schools with both Iraqi and Jordanian children. They then created protection committees and focused attention on specific issues that children prioritised. For SC, the results of these participatory processes – the Unity Circle coupled with the Child-to-Child project –proved that they were engaging on the right issues.

The former director of SC in Jordan told the authors that the Unity Circle was their country offices first 'child rights' project, an observation repeated by the executive director of SC (personal communication between SC director Dennis Walto and IICRD researcher Martha Nelems.). The characterisation of the project as a 'rights' project was largely due to its participatory content, with children engaged as active partners in the research process.

It is unclear what impact involvement in a participatory action research process had on UNICEF's commitment to participatory development. However, UNICEF's emergency education officer in Jordan considers the participatory approaches adopted in the project to be '*interesting and important*' (Deema Jarrar, Emergency Education Officer, UNICEF; Interview) and she personally recommends the Unity Circle methodology to other colleagues at UNICEF as a model for hearing new voices, and as a means of complementing or validating information generated through other research initiatives (Deema Jarrar, Emergency Education Officer, UNICEF; Interview).

Obstacles to organisations acting on project recommendations

All of the key players involved in the Unity Circle project reported a significant improvement in their understanding of the local context of Iraqi children's lives and in their ability to respond to this reality in practical ways and, subsequently, to improve their programme effectiveness. However, there were many obstacles that prevented organisations from responding fully to the project recommendations and incorporating the learning from participatory research into their work with Iraqi children. These obstacles include: organisational inflexibility; a lack of staff continuity within organisations; institutionalised power relationships between the Jordanian government and members of the donor community; the slow pace of social change; and the inability of project partners to influence the discussion about Iraqi children within the broader donor community in Jordan.

Organisational inflexibility

Organisations such as CIDA and UNHCR were operating in a highly politicised environment with rigid bureaucratic processes and predetermined programming approaches and priorities, leaving little room to implement the Unity Circle project recommendations.

The Canadian government, via CIDA headquarters, provided the bulk of the funding for the project and CIDA staff in the Canadian embassy in Jordan and at headquarters in Ottawa actively encouraged the Unity Circle project leaders. The Canadian embassy also organised a high profile launch of the project report in Amman. However, the lack of a programmatic focus on Iraqi refugees in CIDA's Jordan portfolio meant that no new programming was developed to follow up on the project's recommendations. CIDA already had programmatic priorities in Jordan and institutional inflexibility meant that they stuck to them.

UNHCR was the biggest funder of INGOs working with Iraqi refugees in Jordan. In their 2009 umbrella proposal to UNHCR for their work with Iraqi refugees in Jordan, SC included participatory activities with children and training of volunteers and teachers on participatory approaches, in keeping with Unity Circle recommendations. However, a UNHCR staffer asked SC to take the participatory activities out of their proposal. According to SC's former deputy director, UNHCR's insistence on removing this component of SC's proposal was a result both of the expense involved in supporting participatory work with refugees, and also of the new donor environment where UNHCR was under greater scrutiny for the use of its funds. UNHCR was already working in a highly politicised environment, but with growing questions about Iraqi refugee numbers in Jordan compared to neighbouring countries, they appeared unwilling to take risks in supporting programming that was not 'traditional emergency work'. Within the global context of emergency work, participatory approaches with refugees are rarely prioritised, despite the growing consensus that participation of local communities should be a critical component of humanitarian action.[6]

Although SC's former deputy director argues that UNICEF could have '*taken more of the [participatory] approach to heart within their country office*' (Jason Erb, Former Deputy Director for Emergency Programmes, Save the Children; Interview), he acknowledges how fortunate it was that UNICEF allowed them to adapt their programming, by adding participatory activities to their work in Jordanian schools. Given their organisational mandate to support children and after their involvement in the Unity Circle project, UNICEF was better placed than UNHCR to appreciate the importance of ongoing participatory approaches with Iraqi children to understand their evolving experiences in Jordan.

Four years later, in 2012, UNHCR remains a big funder of initiatives for Iraqi families who have remained in Jordan. UNICEF also continues to support Iraqis, though to a much lesser degree, and primarily through the Government, national non-governmental organisations (NNGOs) and community based organisations (CBOs). Although SC US no longer has a budget for Iraqi children, it is soon merging with SC Jordan which has integrated programming for both Jordanian and Iraqi children. RI-SOL was closed down early in 2011, though UNICEF continues to fund the work to support Iraqis of many of the CBOs they had partnered with.

Organisational churn

The rapid turnover of staff within INGOs and UN agencies, especially in contexts considered to be emergencies or crisis-oriented in nature, also contributed to reduced impact of the Unity Circle findings at an institutional level. In all of the participating organisations, all but one of the key staff members involved in the project have left their organisations, and some left very soon after project completion in 2008. According to RI-SOL's former director,

> *I doubt it [Unity Circle] changed many things at the institutional level. No one who worked on that project is in RI-SOL anymore. It was wasted (Angelita Carreda, Former Country Director, RI-SOL; Interview).*

For many of the interviewees, this is a key factor in the lack of long-term change at the institutional level in any crisis situation. Indeed, there is an ongoing debate within the international humanitarian assistance community about how the quick rotation of staff adversely affects organisational learning and the institutionalisation of good practices.

Institutionalised power structures

Institutionalised power structures, specifically the power of the Jordanian government in relation to the donor community, also prevented donors from responding fully to the project's insights. In the lead up to the public launch of the project report, all of the players involved in the Unity Circle repeatedly expressed concern about how the Jordanian government would react to the criticisms levelled against Jordanians, especially the accusation of widespread violence and discrimination against Iraqi children in Jordanian schools. The Canadian government was particularly worried about how the Jordanian government would receive the report. At the launch event, UNICEF Jordan's representative asked the IICRD representative to refrain from directly quoting Iraqi children who talked about violence against them and encouraged instead references to cooperation between Iraqis and Jordanians.[7]

For one of the Unity Circle partners, the results of the report should have been publicised more widely and more effort should have been made to advocate with the Jordanian government on specific problems facing Iraqi girls and boys. RI-SOL senior staff regret that the Canadian government, as the major financial backer of the project, failed to use the release of the Unity Circle report as an opportunity to engage in greater advocacy with the Jordanian government on behalf of Iraqis. RI-SOL's former deputy director argues:

> *If the Canadian government had been more outspoken and in coordination with other government donors had organised strategic releases of the report – with the United States government for example – that might have given us [access to] higher level decision makers. I'm not sure what the Jordanian government would have done, but it would have been useful ... It would have been interesting to see what happened (Angelita Carreda, Former Country Director, RI-SOL; Interview).*

For senior staff at RI-SOL, it is clear that the fear factor held back not just the Canadian government, but all of the key players including themselves:

> *The atmosphere of those days was fear of engaging too much with the government: we feared retaliation and we would be shut down. Instead, what could be done was to work*

at the grassroots level and get in touch with the community (Angelita Carreda, Former Country Director, RI-SOL; Interview).

The slow pace of social change

Not all of the players involved in the Unity Circle project are convinced that the balance of power between host country and donor partners was a critical issue in project follow-up. For SC's former deputy director, the problem instead was the strength of societal or cultural norms in Jordan, where corporal punishment in schools is widely accepted. SC had, for example, first-hand experience with parents who wanted teachers to beat their children if they didn't perform well at school. For the former deputy director, if there was a failure on SC's part, it was not the lack of advocacy with the Jordanian government but instead the lack of recognition for how lengthy the process of changing social norms can be:

The pace and degree of social and cultural change takes a long time on these issues. It's very slow and we were hoping for more rapid change of pace than what was feasible in Jordan (Jason Erb, Former Deputy Director of Emergency Programmes, Save the Children; Interview).

Failure to consolidate the learning about Iraqi children within the donor community

From the authors' perspective, perhaps the most significant failing in follow-up to the Unity Circle was that the learning from the project about Iraqi children was not consolidated among the wider donor community in Jordan. The data about Iraqi children was not used by other organisations as a baseline from which to update information. Nor did the inter-agency dialogue about programming strategies within the Iraqi community mature.

As one of the few senior project participants still working with Iraqi refugees in Jordan, the former deputy of RI-SOL claims that the Unity Circle report did not successfully shift the discussion about Iraqi refugees within the donor community.[8] She argues the following:

My constant issue remains that it was a wonderful attempt to create understanding but, looking back today, the same conversations are taking place with UN agencies and donor organisations. They are asking the same questions and saying 'what are our strategies to increase protection'? Questions are repetitive, and fledgling organisations have grown up, but at the same time, this report has been lost. I reminded UNICEF at this meeting about the Unity Circle report. I said, 'Why is the report not mentioned? And subsequent reports? Why are we asking the same questions? Why don't we instead update the information?' It shows that we collect information, but it impacts at the individual not the institutional level (Karishma Budhdev, Former Deputy Director, RI-SOL; Interview).

Thus, despite initial enthusiasm about the Unity Circle project, the donor community was not able to use this research to deepen and enrich their inter-agency dialogue about policy and programming for Iraqi children in Jordan.

Conclusion

The Unity Circle's participatory action research with Iraqi children and their families had an immediate impact on the organisations involved. A better understanding of the local context and lived experiences of Iraqi girls and boys resulted in numerous changes to how organisations worked and the nature of their programming. These changes included a commitment to increase public awareness of violence and discrimination against Iraqi children, improved relationships with Iraqi volunteers, confidence to engage in child protection programming, a reprioritising of child protection interventions, a refocusing on systemic causes of threats to children, and a growing commitment to participatory development in general. Equally clear was that organisational inflexibility, rapid staff turnover, institutionalised power structures, cultural constraints and a failure to consolidate findings in the broader donor community weakened the impact of the Unity Circle project in the long term.

Participatory processes designed to involve adults in the development of the programmes and policies intended to help them are gaining acceptability as standard best practice in the international development community. Yet organisations' incorporation of learning from children remains woefully inadequate. Nevertheless, they can benefit a great deal from such incorporation.

Organisations need to find new ways of working better to respect the knowledge and experiences of children, and to become more flexible in developing programmes based on this learning and children's stated priorities. Steps in the right direction include a reorientation within organisations towards knowledge retention, and support for longer-term staff positions. A willingness to move outside pre-determined agendas and to respond better to local context and the perspectives of children and families as they adapt over time is important for the effective evolution of the field. An increased commitment to sharing learning across the international development community is crucial, particularly bringing learning about child protection into other areas such as the environment, economy, health, and education. Moving beyond limited notions of child protection to explore the systemic causes of abuse, neglect, and exploitation of children, and violence against them, is a fundamental step in prevention and early intervention. The insights of children and families gained through participatory processes are critical to enabling underlying issues to be identified and addressed.

Notes

1. For the 2007/08 school year, the Jordanian Ministry of Education reported 20,600 Iraqi students enrolled in public schools and 4,000 in private schools.
2. The authors interviewed the following project participants via Skype calls during September and October, 2011.
3. Although the research process was led by IICRD team members, Jordanian and Iraqi adults were trained as researchers and led sessions with children and families whenever possible.
4. The authors interviewed the following: Karishma Budhdev, former deputy director, RI-SOL; Angelita Caredda, former country director, RI-SOL; Jason Erb, former deputy director for emergency programs, SC; Nadine Haddad, child protection specialist, SC; and Deema Jarrar, emergency education officer, UNICEF. CIDA staff declined interviews.
5. Researchers observed that in families where sons had been killed in Iraq, boys were treated in the same highly protective manner as girls.
6. For example, participation is one of the core principles of the Inter-Agency Standing Committee Guidelines on Mental Health and Psychosocial Support in Emergency Settings.
7. As reported by one of the IICRD researchers and authors of this article.
8. The former deputy director of RI-SOL still participates in the education coordination group, but on behalf of a different INGO.

Concluding reflections: how might we really protect children?

William Myers and Michael Bourdillon

The articles contained in this issue present a panorama of good protective intentions gone awry. International or national policies presumed to benefit children appeared in field research to be ineffective or counterproductive for the children. Depending on the situation, the problem could primarily be one of poor interpretation or implementation of a child protection policy. This is a management issue, which can be resolved by competent governance. National governments and international development programmes know how to handle such issues; the question is one of will and priority. However, in some cases, perhaps even in most cases, it appears that at least part of the underlying dysfunction was with the policies themselves, some of which might be misguided. This is a far more serious problem, since it suggests that those making the policies may know rather less about child protection – what is needed and how to provide it – than has been assumed.

As we pointed out in the Introduction, much has been learnt and put to good use about how to protect the lives, well-being, and development of very young children. But that is not equally true for children from young middle childhood onwards. These children, especially in poor countries, have been the primary subjects of concern in the papers in this journal issue. We end by briefly reflecting on them as a group, taking into consideration pertinent thought from elsewhere, to see what lessons we might draw and consider to helpfully inform development theory and practice.

Lessons for consideration

The protection of children involves much more than keeping them safe; our objective should always be to promote their well-being and development.

The papers in this collection indicate that conceiving of 'protection' too narrowly creates serious problems for children, an observation that has become common among both field practitioners and specialists. The problem is widespread, and it may have an institutional aetiology that should concern development professionals. They may have the matter wrongly framed.

Experts and researchers have lately been asking whether a current organisational tendency to define 'child protection' as a discrete policy or programme area, separated from health, education, and other development-oriented activities, may be encouraging single-issue approaches that by addressing a particular ill in isolation only generate new ones in different areas of children's lives. We believe there is growing informal agreement that protective activities need to approach children more holistically and contextually, in a much broader perspective than is now the rule. Articles in the last section of this collection explore what is involved in taking and implementing a more comprehensive view, and even these modest illustrations suggest that the task is challenging. They imply that much about common practice needs to be thought and done differently.

What must development practitioners, planners, and policymakers do to provide for necessary protection of children within more comprehensive but still realistic institutional frameworks? Much experimentation and innovation is still needed to produce effective answers to this question, but two general directions for progress can now be discerned: one has to do with the development of children, and the other with the development of society.

In considering the development of children, we must view them holistically and in their social context, for issues of endangerment and protection may be physical or psychological, and they tend to concentrate at the interface between children and society. For this purpose, biologically-oriented theories of child growth and development, commonly used for health and education purposes, may have too limited a view of children's makeup and of their interactions with their social context. An alternative and potentially more useful view pays more attention to the cultural foundations of human development. This kind of approach appreciates that children are not passive recipients of culture and community, but reach out to learn from mentors, develop identities, and join community through their transactions with adults and other children, acquiring a broad range of attachments and cognitive skills necessary for survival and development (Rogoff 1991, 2003). Such an approach takes into account children's strengths and initiatives as well as their weaknesses and vulnerabilities, an important perspective for thinking about risk, resilience, and protection as transactions between children and others in their environment.

Furthermore, children's individual development is tied to that of their society. Current national and international development strategy seeks to improve people's lives by generating economic growth. Such strategy assumes that people progress because they and their society become richer and better able to purchase necessities and amenities. Within this approach, human development is interpreted as 'human capital', and its ability to contribute to economic growth, primarily through industrialisation. In this this line of thinking, the protection of children may not appear as an important development factor because it does not contribute directly to economic growth; it appears to be a consumption expense more than an investment.

For environmental and other reasons, it is increasingly recognised that a development paradigm based on perpetual economic growth is probably not sustainable over the long term, and must be replaced by one that is. A leading contender as an alternative paradigm is a human development and capability approach (Deneulin and Shahani 2009; Nussbaum 2011; Sen 1999, 2010), which focuses on the improvement of quality of life directly, rather than indirectly as a product of economic development. It proposes to create more life choices for people, especially for the poor, and provide them with the skills, social resources, and other capabilities they need to take advantage of these choices. The objective is less the accumulation of riches than a rich life with health, education, social and self-expression, democracy, fairness, and justice. Promoting children's opportunities for choice and providing them with the life capabilities they need would logically be a central concern under this paradigm, as would be protecting their access to those choices and capabilities. Even in emergencies, as when children must be protected from the effects of natural disasters and armed conflicts, the purpose is not just to maintain their physical safety, as important as that is, but to ensure their continuing development of opportunities and capabilities, as children in the present and as adults in the future.

Although this development paradigm provides an excellent conceptual and policy framework for protecting children, the tools for putting it into practice have yet to be forged. Such tools are necessary not only to detect threats to children's well-being and development in everyday life (for example, as consequences of living in poverty), but also for providing protection against patterns and clusters of risk rather than just focusing on single issues, as is often done. The creation of instruments to analyse, plan for, monitor, and evaluate the well-being and holistic development of children as part of a general human development approach is a demanding task that deserves high priority.

The protection of child well-being and development requires that interventions fit children's social and economic context and involve their families and communities.

Ethnocentrism has long been recognised by social scientists as undermining the effectiveness of some universalised international child protection policies. The problem is that it promotes intervention models that seem reasonable to policymakers in the capitals of power but turn out to be irrelevant or counterproductive for children in other places (for example, Boyden 1997; Bourdillon *et al.* 2010). If there is anything that has been learnt from accumulating field studies, it is that standardised, one-size-fits-all policies to protect children do not work very well. De-contextualised policies can harm children, as some of the articles in this volume illustrate. The development community needs to own up to this problem and address it.

Community participation has for many years been widely accepted as an effective check on ethnocentrism, a hallmark of good development practice, and essential for a wide variety of projects to succeed. Strategies developed from a community's own perceptions of the problems it faces and of its ability to overcome them are more likely to succeed than are strategies imposed by outside authorities. Necessary skills have been developed for such community involvement and capable personnel are widely available through NGOs. What seems to be lacking is the understanding at the policy level that the same kinds of participative dynamics and community buy-in are needed for effective protection of children.

Effective protection of child well-being and development usually needs to be contextualised to the local or regional level to fit the physical and social contexts in which children live. Contextualisation involves engaging families and communities, which are usually the primary defenders of children. There are many schemes and tools available for doing this – some project examples appear in the final section of this volume. It has, however, been suggested that some child protection policies are so attached to international legal norms and a law-based approach that policymakers, donors, and other decision makers simply forget to apply what they have learned elsewhere. A basic legal foundation for children's protection is important, but the assumption that law and enforcement should be the primary structure through which the protection of children operates is highly debatable. The studies presented here suggest an urgent need for a shift of emphasis from law and normative standards, to community and social relationships as the centre of protective effort.

There have been a number of attempts to bring legal and community perspectives together in creative partnership. One of the most audacious and interesting has been in Brazil, which established in law both a national statute to provide a policy framework for the protection of children, and a system of local councils to implement it. These councils comprise equal numbers of representatives from local government and civil society, who are charged to work together in overseeing the protection of children in their area (Boyden and Myers 1995: 11–29). The outcomes for children of this intriguing experiment, which may be important to learn from, have yet to be made known.

In interventions for children, the interests of children must explicitly be a primary consideration.

In the case studies in this issue, what was best for children seemed too often to get lost, whether from neglect, ignorance, or displacement by adult political or institutional interests. That is a human rights violation. Such disregard for children's welfare even in actions ostensibly for their protection is so common and debilitating that the drafters of the UN Convention on the Rights of the Child made a point of requiring that, '*in all actions concerning children ... the best interests of the child shall be a primary consideration*' (UN 1990: Article 3.1). A 'primary consideration' is not necessarily the only or overriding consideration: there is no *a priori* reason why the interests of a particular child should override the fundamental interests of family, the community, or the state. Indeed, as we pointed out in the Introduction, the survival, well-being, and development of children advances with the economic and social good fortune of the society to which they belong. But the best interests of the child should be at least of very high priority, and nudging priorities of protective policy and practice just that far would revolutionise the situation of vulnerable children for the better. This step forward is within the purview and reach of development policy.

A practical place to start is in policy formulation, programme and project planning, and evaluation. It is now recognised good practice to include social impact analysis, along with environmental impact analysis, in the design and assessment of programmes and projects. Where children comprise roughly half the population, which is the case in much of the developing world, it would be reasonable to devote roughly half the attention to protecting and promoting their well-being and development. In the context of a human development paradigm, this focus on children would be logically essential.

Some of the case studies in this volume indicate that organisational structure and procedures to promote child protection merit scrutiny and probably reform. These studies, like evidence from elsewhere, suggest that the protection of children may not be easily amenable to bureaucratisation. Attempts to organise child protection as a public service like health and education have produced unsatisfactory experiences in both rich and poor countries. There appears to be something inherently relational and situational about the protection of children that resists universalisation and standardisation, which is one reason for growing interest in placing more activity related to protection within the context of community. The case studies presented here strongly suggest that families and communities typically do care about the best interests of children in their care, even when the ways they define and serve those best interests do not necessarily conform to universalised ideas and policies coming from government or elsewhere. There clearly is a need to better understand how families and communities regard the well-being and development of their children before jumping in with external interventions that can make already difficult situations for children even worse.

All protective policies, programmes, and activities for children should be accountable according to their observed outcomes in affected children's well-being and development.

One way of making sure that children's interests receive primary consideration is to ensure that programmes and interventions are accountable to the well-being of children. The development community has well-established demands of accountability to intended beneficiaries, including posing the right questions, identifying the problems correctly, applying remedies to fit specific situations, and monitoring and assessing outcomes. These evidence-based demands call for adequate research prior to, during, and after intervention, a good practice that should be extended more rigorously to child protection policies and programmes, where it has been notoriously lacking. For instance, the longest-standing international child protection policy – universal legal minimum age requirements from 1919 onward for children to enter work – has never been properly evaluated to empirically determine its impact on children's well-being, although there is increasing case evidence that it harms many children (Bourdillon *et al.* 2009). This is despite the fact that good governance practice calls for routine impact analysis of all major social policies.

Accountability depends on institutionalisation of a self-critical attitude often lacking in child protection programmes, which in our experience are too often inclined toward unwarranted arrogance and dismissal of even the most justified criticism, often arising from an assumption – again unwarranted – of holding superior knowledge and values. Through insistence on management by results, good development programmes train both public and civil society organisations to cultivate habits of continual assessment of progress against outcomes. This mentality should be more firmly encouraged in regards to child protection activities. It is crucial, however, that outcomes be defined in terms of children's well-being and development (such as health and nutrition status, or literacy level) rather than merely in terms of programme through-put (such as health services received, or years of schooling). Time frames should also realistically reflect the time needed for expected results to appear.

We have pointed out that outcomes for any intervention concerned with child well-being must go beyond the immediate physical and psychosocial safety of children. They should also provide children with the ability as adults to:

- live independently;
- fit constructively into society; and
- establish the kind of relationships that allow family life.

These are very long-term outcomes: they need to be assessed through research, using longitudinal or other monitoring and evaluation techniques that provide insight into details of the impact of interventions on the lives of children and of their communities over time. This certainly involves much more than routine statistical assessments of conformity to standardised norms. It calls for continued engagement with communities and at least samples of children as they grow up. Good accountability practice suggests that both children and communities should have a voice in designing and overseeing such time-sensitive evaluative research.

Evaluative studies should provide information about outcomes needed to evaluate policies, legal norms, and implementation guidelines for the protection of children in order to ensure that:

- policies, norms, and guidelines are needed and correctly defined;
- they are correctly interpreted; and
- they are properly applied to the benefit of children.

These judgments can be made only if protective activities are assessed in terms of outcomes in the well-being and development of the children involved. Only those programmes demonstrating an increase in child well-being and development should be considered successful and deserving of funding and political support.

To ensure that protective policies and programmes are beneficial to children, it is necessary to consult them and to take their opinions seriously.

The development community has long recognised that engagement with the beneficiaries of programmes and projects, starting from the planning stage, is more likely to produce benefits than are predetermined packaged activities imposed from outside. It should come as no surprise that this is true of programmes to protect children. The situations and concerns of children are better understood by children and their communities than by outsiders, and, as articles in this volume illustrate, children can often shed light on the impact of successful or failing programmes with details unlikely to be uncovered from anywhere else. Making use of this insight, promising examples of protecting children through community-based interventions in which children participate have become increasingly common. The studies in this collection graphically illustrate how input from children can improve our understanding of their situation and the problems they face, which enables more accurate design, targeting, and prioritisation of interventions.

Sometimes the perceptions of children differ from those of adults in their families and communities. Children experience problems differently and may have concerns about their status and future prospects that differ from the views of adults around them. This is often the case, for instance, in regards to gender roles. Developmental psychologists (for example, Rogoff, as cited above) point out that children actively make choices for pursuing their own development, based on their aspirations and interests. What lies behind the choices can be learned only from the children. Since adults in the family and community may see the issues quite differently, the question arises: how should we hear and weigh both? Good practice certainly calls for listening to adult observations and opinions as well as children's, but adults cannot be assumed to accurately represent their children's views.

Since the UN Convention on the Rights of the Child requires that in all matters concerning the child, his or her opinions be heard and given due weight (UN 1990: Article 12), consultation with children has become widespread through the work of many child defence organisations. More than that, active participation of children has expanded beyond mere consultation into taking initiatives of their own to further their own interests. As we write, a committee of the Norwegian Parliament has nominated three child defence organisations – UNICEF, Save the Children, and Concerned for Working Children – to receive the 2012 Nobel Peace Prize for their work in promoting children's participation.

Many years ago, a leading expert on child protection pointed out that '*the main resource in any project is the children themselves*' (Ennew 1994: 7). There are many ways in which this resource can be fostered and utilised, and some are illustrated in the final section of this collection. One of the most interesting and promising is the formation of children's own organisations, controlled by children but with adult support. In such organisations, children have the chance to work not only for their own protection and development, but also for that of other children and the betterment of their societies. For example, associations of working children have become common in parts of Latin America, South Asia, and Africa, and some of them have influenced government and NGO policies (Liebel 2012: 199-225).

The participation of children in decisions that affect them has been shown to augment their confidence and growing competence (Lansdown 2005). A variety of techniques have been developed to encourage and help children to their express their ideas and feelings about issues that concern them, and children regularly report that such participation is of great value in building their confidence and expanding their horizons. It is good policy to encourage and facilitate participation of children not only as a human right and for what they can achieve through it to protect their own well-being and that of others, but also as a powerful instrument of child development in its own right. Especially in development policy centred on human development, child participation should be an element as central as education. Indeed, as developmental psychologist Barbara Rogoff (1991) has pointed out, it *is* education; participation with others in solving problems is one of the most basic and powerful mechanisms of human learning. All children should have access to it.

Parting thoughts

As indicated in the Introduction, there is a nagging worry that that 'child protection' as a category of activity is too restricted and reactive, inviting narrow agendas dictated by the incursion of problems to be resisted or reduced rather than by opportunities to be seized. A matter of some discussion these days is whether that notion should just be discarded, leaving us with the sole objective of pursuing children's well-being and development in all situations, some of which are admittedly more difficult than others, but all of which have possibilities for ensuring that children are healthy, learning, engaging with society, and otherwise progressing. This suggests that there are healthy growth opportunities even in adversity and stress; indeed, some human development literature makes that very point, including for the development of children (e.g., Aldwin 2007: 271–340).

In light of the lessons drawn from this collection of articles, we propose some thoughts and directions for the future of protecting children. We should certainly consider approaching adverse, even dangerous, conditions for children as calls to innovate in guaranteeing their holistic well-being and development in the context of social support. We could replace the concept of 'protection' with the the idea of promoting well-being and development, drawing on assets that may inhere in the situation. The greatest of those assets is children themselves, and we should actively engage them. We need to consider what we can do for and with children, and how we, in partnership with children, can improve their position in the context of society.

If we routinely work more directly for children's well-being and development in *all* situations, there might be less need for palliative humanitarian programmes, and there might be more incentive to confront and attend to the causes of dangers to children. Rather than dividing children's experiences into the categories of 'risk' and of 'protection' (or 'resilience') against that risk, we might seek the seeds of personal and social transformation that lie even within adversity: work with and for children can seek holistically to prepare and assist them to thrive and develop through life's difficulties as well as its rewards. To do this, we need to make our societies more just and empathetic. Such realignment of emphasis and direction might profoundly change the face of development policy and practice.

References

Aldwin, C. M. (2007) *Stress, Coping and Development: An Integrative Perspective*, New York, NY, USA: The Guilford Press.

Bourdillon, M., D. Levison, W. Myers, and B. White (2010) *Rights and Wrongs of Children's Work*, London, UK: Rutgers University Press.

Bourdillon, M., W. Myers, and B. White (2009) 'Reassessing working children and minimum-age standards', *International Journal of Sociology and Social Policy* 29 (3/4): 106–17.

Boyden, J. (1997) 'Childhood and policy makers: a comparative perspective on the globalisation of childhood', in A. James and A. Prout (eds.) *Constructing and Reconstructing Childhood* (2nd ed.), London, UK: Falmer Press.

Boyden, J. and W. Myers (1995) 'Exploring Alternative Approaches to Combatting Child Labour: Case Studies from Developing Countries', Innocenti Occasional Papers, Child Rights Series Number 8, Florence, Italy: UNICEF.

Deneulin, S. and L. Shahani (eds.) (2009) *An Introduction to the Human Development and Capability Approach: Freedom and Agency*, London, UK: Earthscan.

Ennew, J. (1994) *Street and Working Children: A Guide to Planning*, London, UK: Save the Children.

Lansdown, G. (2005) *The Evolving Capacities of the Child*, Florence, Italy: UNICEF, Save the Children.

Liebel, M. (2012) *Children's Rights from Below: Cross Cultural Perspectives*, London, UK: Palgrave.

Nussbaum, M. (2011) *Creating Capabilities: The Human Development Approach*, Cambridge MA, USA: Harvard University Press.

Rogoff, B. (1991) *Apprenticeship in Thinking: Cognitive Development in Social Context*, New York, NY, USA: Oxford University Press.

Rogoff, B. (2003) *The Cultural Nature of Human Development*, New York, NY, USA: Oxford University Press.

Sen, A. (1999) *Development as Freedom*, Oxford, UK: Oxford University Press.

Sen, A. (2010) *The Idea of Justice*, London, UK: Penguin.

UN (United Nations) (1990) 'Convention on the Rights of the Child', available at http://www.unicef.org/crc/ (retrieved 29 February 2012).

Index

Page numbers in **bold** refer to tables.
Page numbers in *italic*.
Page numbers annotated with 'n' refer to chapter notes.